BASIC Stamp

BASIC Stamp

Dr. Claus Kühnel
Dr. Klaus Zahnert

Newnes
Boston Oxford Johannesburg Melbourne New Delhi Singapore

Library of Congress Cataloging-in-Publication Data

Kühnel, Claus, 1951–
   BASIC stamp / Claus Kühnel, Klaus Zahnert.
      p.   cm.
   Includes index.
   ISBN 0–7506–9891–8
   1. Programmable controllers.   I. Zahnert. Klaus, 1936–   .
   II. Title.
   TJ223.P76K84   1997
   629.8'9—dc21
                                                           96–46695
                                                           CIP

British Library Cataloguing-in-Publication Data
A catalogue record for this book is available from the British Library

The publisher offers special discounts on bulk orders of this book.
For information, please contact:
Manager of Special Sales
Butterworth–Heinemann
313 Washington Street
Newton, MA 02158-1626
Tel: 617-928-2500
Fax: 617-928-2620

For information on all Newnes electronics publications available, contact our World Wide Web home page at: http://www.bh.com

10 9 8 7 6 5 4 3 2

Printed in the United States of America

Contents

Preface

Parallax, Inc. is an eight-year old company located in Rocklin, California. Parallax started with the development and production of memory cards for the Apple IIgs.

While Parallax was working on alternative development tools for Microchip's PIC controller in 1991, the first ideas for the BASIC Stamp were born. The BASIC Stamp idea was to implement a token interpreter in the program memory of the PIC and to add a serial EEPROM as token and data memory. With a minimum of additional hardware for clock and reset, the whole microcontroller was ready.

To generate the tokens for the BASIC Stamp, Parallax designed a DOS-based development system. This development system includes Editor, Compiler, and Downloader in one environment. Because there are some debugging features, a quick cycle for editing, compiling, downloading, and testing is possible.

After sales of about 50,000 BASIC Stamps in fifteen countries in the last three years, experience and customer feedback gave Parallax the idea for a new BASIC Stamp—the BASIC Stamp II (BS2).

Problems for some computers with the direct-driven serial interface via the parallel port led to a truly serial interface via a COM port. The instruction set of P(arallax)BASIC was extended and some instructions were enhanced with advantageous features.

This whole time, Parallax was also working on the technological design of BASIC Stamps. For the BASIC Stamp, there was a conventional design solution in the BASIC Stamp Single-Board Computer. The BASIC Stamp chip set was placed on a 1.5 × 2.5-inch printed circuit board, which included contacts for a 9-V battery 6F22, PC and I/O headers, and a prototyping area that provides space signals and extra parts.

Today the carrier board is fully passive, consisting of battery clips, PC and I/O headers, and an adaptor (14 pins) connecting the new BS1-IC. The BS1-IC is a very compact module perfect for use in custom designs. Because the modules have a necessary circuitry on-board, the customer design only needs to provide application-specific circuitry.

The same design solution was chosen for the design of BS2-IC. BS2 is the BASIC Stamp microcontroller with the extended and enhanced instruction set and the new development system, but the same successful philosophy.

Introduction to Microcontrollers   *1*

This book describes two microcontrollers from Parallax, Inc., BASIC Stamp I and II, as well as some applications. Both BASIC Stamps are suited for many types of users. BASIC Stamp users cover a wide range—from electrical engineers to technicians who need tools for measurement and control, to students on different levels, and to hobbyists or beginners in electronics. In addition, the educational sector is very much impressed with BASIC Stamp.

Before describing the features of a microcontroller, we need to make clear what a microcontroller is. While a professional should be on familiar ground, a hobbyist or a beginner would need some clarification.

This chapter is dedicated to the last type of readers. The professional can start with the next chapter. After reading this chapter, the beginner knows the terms and can better understand the following chapters.

1.1 Function Groups of a Simple Microcontroller

To explain the function groups of a microcontroller we want to compare it with the well-known personal computer (PC). In Figure 1.1 the PC is shown in a desktop version. Each personal computer is built using more or less the same components. The keyboard and mouse (mouse not shown in Figure 1.1) serve for input, and the video graphics array (VGA) monitor outputs the results in a graphic and/or alphanumeric display.

The mainboard contains the central processing unit (CPU), the internal memory (RAM), the interfaces to the serial and parallel ports, and the hard disk, in addition to the floppy disk. Slots are reserved for further enhancements using standardized PC cards.

The task of a PC in general is to run a program selected by the user, using such input means as the keyboard or mouse. The program and data of that

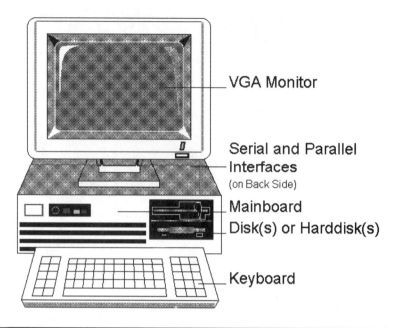

VGA Monitor

Serial and Parallel
Interfaces
(on Back Side)

Mainboard
Disk(s) or Harddisk(s)

Keyboard

Figure 1.1
Desktop PC

application are stored on a hard disk or disk and will be loaded to the PC's memory at run-time. The results of the running program (or queries requiring decisions) are displayed today almost always in graphics mode over a VGA monitor, with the level of resolution determined by the monitor.

Standardized serial and parallel interfaces provide data exchange to external equipment. All PC users know the parallel port is used to connect a printer and the serial port to connect a mouse for data input or another PC for data exchange.

Microcontrollers for industrial applications have another shape. Figures 1.2 and 1.3 show two possible outfits for microcontrollers meant for industrial application.

The industrial control board shown in Figure 1.2 turns BASIC Stamp I into an industrial controller. This board accepts a standard BS1-IC module and provides eight digital I/O connections. These connections may be used for two purposes, namely, protected DC inputs for contact closures and logic levels, and high-current (50 VDC, 1 A) DC outputs, transient protected for inductive loads. In addition, the board has provisions for eight Opto-22 plug-in modules to enhance the process interface.

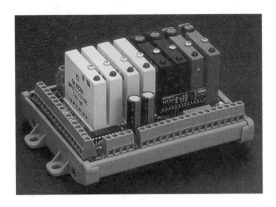

Figure 1.2
Industrial control board

The PC-104 format is used in industrial applications of PC-based components. Figure 1.3 shows the PCStampII I/O board that allows up to four BS2-ICs to serve a master IBM PC. The PC can be a PC-104 single-board PC, or it can be a desktop/laptop system.

We have seen two realizations of microcontroller components for industrial use that look very different. But it is important to ask what a microcontroller

Figure 1.3
PCStampII I/O board

and a PC have in common, for example, our well-known PC and the micro-controller in a washer.

In Figure 1.4 the front of a typical front-loading washing machine is shown. We are interested only in the elements in the user unit.

Our sample washer has three push buttons on the left side and one turning knob for selecting a washing program on the right side. In the middle of the user unit we find three signal lights with different colors. These seven elements all contribute to the dialog between the washer and the user.

Although we have a knob for selecting a washing program, the washer's internal microcontroller knows only one control program. But this control program for the washer knows some parameters, such as temperature and amount of water, time, and revolutions for each step in the washing process, so the user gets the impression of more than one program.

The push buttons on the left influence the entire washing process. One push button serves as a door opener. Other buttons allow the user to reduce the amount of water if the washer is only half-loaded and to spin-dry with two different settings.

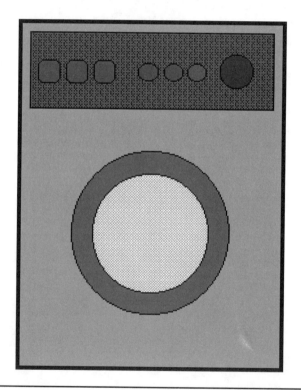

Figure 1.4
Washer

The signal lights display certain states in the washing process. Door-opening is only possible if there is no water in the washer. For some kinds of laundry there must be a stop before spin-dry, because the dry laundry must be removed immediately after. These states are important to the user and can be displayed by the signal lights.

Because the washer's internal microcontroller knows only one program, there is no need for a memory that saves more than one program, as does a hard disk or a disk. The control program for the washer can be stored in an electronic device called EPROM (Electrically Programmable Read Only Memory). An EPROM is a typical program memory. The microcontroller is able to read only the contents of this memory. Programming this memory requires electronic equipment for programming—simply called a programmer.

The washing process knows not only the program; there are data such as temperature and amount of water, as well as other information. To store these data Random Access Memory (RAM) is required. Random access here means writing and reading of data at any time.

The separation of memories for programs (EPROM) and data (RAM) is characteristic of microcontroller applications. The CPU retrieves step-by-step instructions from program memory and, if required, data from data memory. After processing data are stored in data memory (if required) the next instruction can be fetched from program memory.

The CPU and both memories are function prompts of each microcontroller. Figure 1.5 shows the block diagram of a microcontroller.

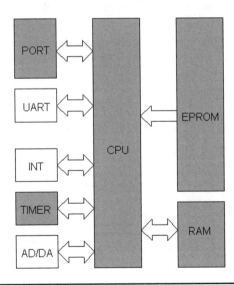

Figure 1.5
Block diagram of microcontroller

To set the program parameters as selected by the buttons, the program has to take into consideration the state of these buttons. For digital input and/or output the microcontroller recognizes ports. A port is built by a certain number of connections between the microcontroller and the process, often a factor times eight.

Data exchange between microcontrollers or between a microcontroller and a terminal or PC happens via serial ports based on the RS232 standard. For effective data handling, serve one or more Universal Asynchronous Receiver Transmitter (UART) type serial communication interfaces. Another possibility is to solve the serial data exchange entirely in the software.

Because the microcontroller is designed for process-related applications with real-time character, some other function groups are implemented in a microcontroller.

Modern microcontrollers have fairly comfortable interrupt systems. An interrupt breaks the running program to process a special routine—called the interrupt service routine. An external event, such as reaching the upper limit of a temperature range, requires an immediate reaction. This event generates an interrupt; the whole system will be frozen, and in the interrupt service routine the heater will be switched off before the system can be damaged. If the microcontroller has to process many interrupts an interrupt controller can be helpful.

To fulfill timing conditions microcontrollers have one or more timers implemented. The function blocks usually work as timer and/or counter subsystems. In the simplest case we have an eight-bit timer register. Its contents will be incremented with each clock cycle (CLK). Any time the value 255 is reached, the register will overflow with the next clock. This overflow is signalized to the CPU. The actual delay depends on the pre-load value. If the pre-load value equals zero, then the overflow will occur after 256 clock periods. If the pre-load value equals 250, then the overflow will occur after six clock periods. A block diagram of a simple timer is shown in Figure 1.6.

Last, but not least, in Figure 1.5 we have an AD/DA subsystem. For adaptions to a real-world process, analog-to-digital and/or digital-to-analog con-

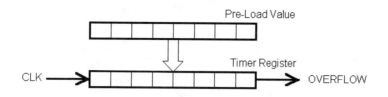

Figure 1.6
Timer

verters are often required. Here is not the place to discuss all features of these rather complex subsystems. A very simple explanation should be enough for this introduction.

To control a process with a microcontroller, we have to measure physical units and change them into values understandable for the microcontroller. The microcontroller knows only bits (1 or 0) or multiples of bits. But the measured process unit is in general an analogous value with theoretical endless resolution. This means that our measured value could have each possible value in the measuring range. The resolution reached depends on the quality of the measurement equipment used.

Figure 1.7 shows the characteristic of a virtual analog-to-digital converter with a resolution of four bits. The input voltage range should be limited to 5 V. This range will be divided into 16 steps on the digital side. Therefore one step is equivalent to a voltage of 0.3125 V. This value is the resolution of the analog-to-digital conversion at the same time. You can see it marked in Figure 1.7; for example, all input voltages between 2.5 V and 2.8125 V generate the same result.

From this consideration it follows that the better the resolution of an analog-to-digital converter, the better the transformation of the converted values to the digital world. Unfortunately it is true that costs also increase with the resolution. So each application requiring analog-to-digital converters needs special consideration of these problems. These explanations are also valid in the opposite case, that of the digital-to-analog converter, so there is no need to address it separately.

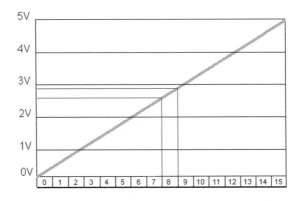

Figure 1.7
Characteristric of an analog-to-digital converter

1.2 Some Important Terms

Here some important terms will be explained for a better understanding in the following chapters.

Internal Architecture of a microcontroller means the concept of building the internal electronics.

Memory is a function block for program and data storage. Here it is important to distinguish between non-volatile and volatile memories. Non-volatile memories are required for storage of a program so the system does not have to be reprogrammed after the power has been off. Working variables and intermediate results need to be stored in a memory that can be written (or programmed) quickly and easily during system operation, a volatile memory. It is not important to know these data after the power is off. Examples of non-volatile memories are EPROM and OTP-ROM. The typical example of a volatile memory is the RAM. There are many more types of memory, but here they are of no further interest.

EPROM describes an Electrical Programmable Read-Only Memory. This memory is programmed by programmer equipment. Memory is erased by irradiation of the chip with ultra-violet light through a crystal window in the ceramic package. EPROMs are typical memories for program storage. Some microcontrollers have EPROM included so the CPU itself is involved in the programming process.

OTP-ROM describes a One-Time Programmable EPROM. This type of EPROM is one-time programmable because the package is without a crystal window and therefore not UV-erasable. Using cheap plastic packages without a window instead of windowed ceramic packages decreases the cost. Therefore the PROM included in a microcontroller is often an OTP-ROM.

EEPROM describes an Electrical Erasable and Programmable Read-Only Memory. For reading, it is no different from a normal EPROM. The program cycle is about 10 ms for a byte or a block of bytes. Due to the program time, the EEPROM is suitable for storage of seldom changing data, such as initialization or configuration data. For modern EEPROMs today 10 million program cycles are possible.

RAM describes a Random Access Memory. That means programming and reading at any time. RAMs are typical memories for data storage and are volatile.

Oscillator describes a circuit that produces a constant frequency square-wave that is used by the computer as a timing or sequencing reference. A microcontroller typically includes all elements of this circuit except the frequency-determining component(s) (crystal, ceramic resonators, or RC components).

Reset Circuit generates a reset-impulse to reset the computer, in some cases. The most important reset is the Power-On Reset. Switching Power-On starts the program of the microcontroller.

I/O Ports are the connections to the process. Mainly the ports are bit-programmable in both directions.

Watchdog means a counter circuit that must be reset by the running program. If the program hangs, no watchdog reset can occur and the watchdog counter overflows. As a result of the overflow, the watchdog initiates a reset and avoids wild running of the microcontroller.

Real-Time Clock/Counter is a counter circuit able to count real-time pulses from the process side or is controlled by a clock generated by the internal clock.

Terminal describes equipment for serial I/O to the microcontroller. In the most cases a PC running a terminal program is used.

Program Installation means the installation of a user program on the hard disk of a personal computer. The installation process includes copying the file(s), uncrypting these when needed, generation of a program group in a Windows environment, and so forth.

Program Initialization means to provide an installed software with the required constants and/or parameters. For example, initialization would provide baud rate and handshake parameters for serial communication.

Instruction Set describes the whole list of instructions that will be understood from the microcontroller. The instruction sets of the two BASIC Stamps differ!

MSB is an abbreviation of Most Significant Bit. In the 8-bit data word D7:D0 = 10101010 the MSB is D7 = 1.

LSB is an abbreviation of Least Significant Bit. In the 8-bit data word D7:D0 = 10101010 the LSB is D0 = 0.

PullUp-Resistor means a resistor that gives a Hi signal in a high-impedance circuit.

PullDown-Resistor means a resistor that gives a Lo signal in a high-impedance circuit.

KByte, MByte are units for bits and bytes. "K" means here not a value of 1,000 and "M" not 1,000,000. In the binary system of the information technique the "K" stands for $2^{10} = 1,024$. "M" stands for $2^{10} * 2^{10} = 1,048,576$.

CMOS (Complementary Metal-Oxide Semiconductor) describes the technology used today in modern microcontrollers. CMOS circuits leave a low power consumption, so they can be used for battery-powered equipment.

CERDIP stands for CERamic Dual In-line Package.

SOIC stands for Small Outline IC package.

Hardware Base of BASIC Stamp I and II

BASIC Stamp I and II are high-level language-programmable microcontrollers with different performances. The performance of the two BASIC Stamps is based on different hardware, i.e., different PIC microcontrollers, and different firmware. In addition, firmware enhancements need more hardware resources, so while the microcontroller PIC16C56 was the hardware base for the BASIC Stamp I, the newer BASIC Stamp II needs the bigger brother PIC16C57. However, the hardware bases of both controllers are the same in many ways. The differences will be shown.

Microchip has the microcontroller family PIC16C5x designed for mid-range and low-end applications. The PIC16C56 and PIC16C57 are members of this family. The fully static-designed CMOS controllers permit high operating speed at low power consumption. The UV-erasable CERDIP-packaged versions are ideal for code development while the cost-effective one-time programmable (OTP) versions are suitable for production in any volume. OTP versions of the PIC16C56 and PIC16C57 in a compact SOIC package build the kernels for BS1-IC and BS2-IC, respectively, shown later.

2.1 Internal Architecture

The PIC16C5x microcontrollers contain EPROM, RAM, I/O, and a CPU on a single chip. The architecture is based on a register file concept with separate buses and memories for program and data (Harvard architecture). Figure 2.1 shows a block diagram of the PIC16C5x microcontroller family.

The on-chip program memory consists of 1,024 words for PIC16C56 and 2,048 words for PIC16C57 12-bit wide EPROM. The 8-bit wide arithmetic logic unit (ALU) contains one temporary working register (W Register). It performs arithmetic and boolean functions between data held in this W Register and any

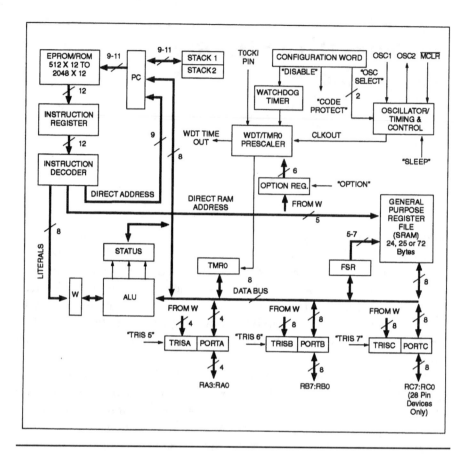

Figure 2.1
PIC16C5x family block diagram. (Reprinted with the permission of the copyright owner, Microchip Technology Incorporated © 1995. All rights reserved.)

file register. The ALU and the Register File are composed of up to 80 addressable 8-bit registers, and the I/O Ports are connected via the 8-bit-wide data bus. Thirty-two bytes of RAM are directly addressable while the access to the remaining bytes works through bank switching.

Figure 2.2 shows the organization of data memory in the PIC16C5x microcontrollers. Programs for the PIC16C5x are typically between 30% and 50% smaller than programs for other 8-bit microcontrollers. Most instructions are single-cycle instructions, and one instruction cycle will take four clock cycles, so the minimum instruction time is 200 ns at a clock frequency of 20 MHz.

A further speed enhancement is provided by overlapping cycles for instruction fetch and for instruction execution. While an instruction is executed, the

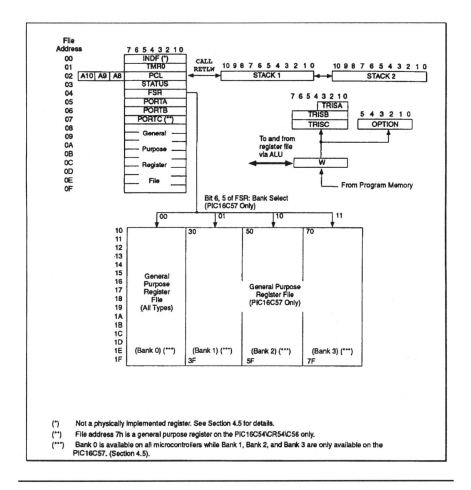

Figure 2.2
PIC16C5x data memory map. (Hints related to original documentation.) (Reprinted with the permission of the copyright owner, Microchip Technology Incorporated © 1995. All rights reserved.)

next instruction is already being fetched from program memory. With this effective instruction handling, PIC microcontrollers reach a performance of 5 MIPS (equals execution of 5 million instructions per second) at a clock frequency of 20 MHz.

The PIC16C5x microcontroller will be offered in several versions, differing by packages, oscillator type, supply voltage, and current consumption. Table 2.1 lists the different versions.

The main functional difference between the versions is the oscillator circuit. This feature will be described in the next chapter.

Table 2.1

Some features of different PIC16C5x microcontrollers

Part number	Erasable	V_{DD}	Maximum I_{DD}	Oscillator	Clock frequency
PIC16C5x-RC	no	3.00–6.25 V	3.3 mA	RC	DC - 4 MHz
PIC16C5x-XT	no	3.00–6.25 V	3.3 mA	XTAL	DC - 4 MHz
PIC16C5x-HS	no	4.50–5.50 V	20 mA	XTAL	DC - 20 MHz
PIC16C5x-LP	no	2.50–6.25 V	32 mA	XTAL	DC - 40 kHz
PIC16C5x-JW	yes	3.00–5.50 V	20 mA	RC, XTAL	DC - 20 MHz

Because all the PIC16C5x have a static design, the clock frequency can be reduced until DC. Applications optimized for very small power consumption may need this possibility. The run of a program can be stopped by halting the oscillator. After a restart of the oscillator, the program starts with the next instruction execution. All registers remain unchanged.

The PIC16C5x-JW possesses a CERDIP package with a quartz glass window, and is therefore the only UV-erasable circuit. These circuits with expensive packages were only used during the period of development. If the application is ready, the less expensive OTP versions of Microchip's microcontrollers are generally used.

In Table 2.1 two PIC16C5x are shaded. The PIC16C56-XT and PIC16C57-HS controllers are the bases for the BASIC Stamp in the BS1-IC and BS2-IC versions, respectively.

2.2 Peripheral Functions

In addition to the CPU capabilities for microcontrollers, the integrated controller functions, such as timer, ports, clock generation, etc., are very important.

2.2.1 Oscillator

The PIC16C56-XT and PIC16C57-HS need a crystal or ceramic resonator connected to OSC1 and OSC2 pins for clock generation. Figure 2.3 shows the required oscillator circuit.

The resistor RS may be required to avoid overdriving for certain types of crystals. The values for the capacitors C1 and C2 are listed in Table 2.2. Higher

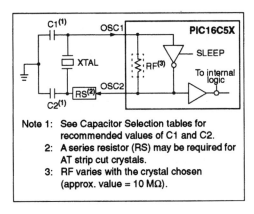

Figure 2.3
Oscillator circuit. (Reprinted with the permission of the copyright owner, Microchip Technology Incorporated © 1995. All rights reserved.)

capacitances increase the stability of oscillation, but also increase the start-up time. The crystal manufacturer should be asked for the recommended values of external circuitry.

If the application circuit contains a clock signal suitable for the PIC16C5x, it can be connected to the OSC1 pin. The OSC2 pin remains open in this case. Figure 2.4 shows the details.

Table 2.2
Capacitor selection for oscillator circuit

Resonator	Type	Clock frequency	C1	C2
Ceramic resonator	XT	455 kHz	150–330 pF	150–330 pF
		2 MHz	20–330 pF	20–330 pF
		4 MHz	20–330 pF	20–330 pF
	HS	8 MHz	20–200 pF	20–200 pF
Crystal	XT	100 kHz	15–30 pF	200–300 pF
		200 kHz	15–30 pF	100–200 pF
		455 kHz	15–30 pF	15–100 pF
		1 MHz	15–30 pF	15–30 pF
		2 MHz	15 pF	15 pF
		4 MHz	15 pF	15 pF
	HS	4 MHz	15 pF	15 pF
		8 MHz	15 pF	15 pF
		20 MHz	15 pF	15 pF

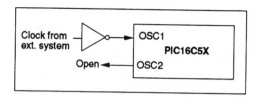

Figure 2.4

External clock input operation. (Reprinted with the permission of the copyright owner, Microchip Technology Incorporated © 1995. All rights reserved.)

2.2.2 Reset

The PIC16C5x has an on-chip Power-On Reset (POR) circuitry, which provides internal chip reset for most power-up situations. The power-on reset is closely related to the oscillator start-up timer circuit, which avoids unstabilities of oscillation in the start-up phase. The oscillator start-up timer keeps the device in reset state over 18 ms after the voltage at pin /MCLR (master clear) has reached its Hi level. In this time the oscillator may start up and must reach stable oscillations.

The power-on reset circuit works reliably if the rate of rise of the supply voltage V_{DD} is no slower than 50 V/s. Also, the voltage V_{DD} must start from 0 V.

In practical applications it happens that, if the changes of the supply voltage are higher than or equal to the specified value, the oscillator circuit is prepared for stable oscillations in the start-up period of 18 ms.

Both problems can be solved with an external power-on reset circuit. Figure 2.5 shows the elements needed.

Resistor R and capacitor C delay the rise of the voltage at pin /MCLR. The resulting delay between a rise of V_{DD} and the reached Hi level at pin /MCLR can be calculated as follows:

$$t_{delay} \approx 1.9 \cdot RC$$

If a delay time of 100 ms is required, for example, then the capacitor has a minimum value of 1.3 µF at a resistor of 40 kΩ. If the supply voltage V_{DD} drops, then the diode D helps to discharge the capacitor C quickly.

Reducing the supply voltage below the specified range can damage the contents of internal registers, and a crash of the whole microcontroller system follows. A brown-out circuit can reset the whole microcontroller in this situation, preventing a wildly-running microcontroller. Generally a brown-

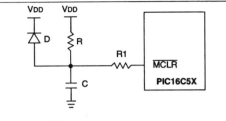

- External Power-On Reset circuit is required only if VDD power-up is too slow. The diode D helps discharge the capacitor quickly when VDD powers down.
- R < 40 kΩ is recommended to make sure that voltage drop across R does not exceed 0.2V (max leakage current spec on MCLR/VPP pin is 5 µA). A larger voltage drop will degrade VIH level on MCLR/VPP pin.
- R1 = 100Ω to 1 kΩ will limit any current flowing into MCLR from external capacitor C in the event of MCLR pin breakdown due to ESD or EOS.

Figure 2.5
External power on reset circuit. (Reprinted with the permission of the copyright owner, Microchip Technology Incorporated © 1995. All rights reserved.)

out circuit is also useful if the power supply rises too slowly to trigger the internal reset, which may happen if a big filter capacitor is in parallel to the power supply.

Figures 2.6 and 2.7 show two brown-out circuits that generate a reset at defined voltage levels.

In Figure 2.6 a Z-diode gives a voltage reference. If the value of the supply voltage V_{DD} is under $V_Z + V_{BE}$ the transistor will shut off and initiate a reset at pin /MCLR.

Figure 2.7 shows the least expensive solution for brown-out protection. The "voltage reference" is built by a voltage divider. In this case a reset will be generated for supply voltages

$$V_{DD} \leq \frac{R_1 + R_2}{R_1} \cdot V_{BE}$$

The following resistor values are usual:

$$R1 = 470 \text{ k}\Omega; R2 = R3 = 2.2 \text{ M}\Omega$$

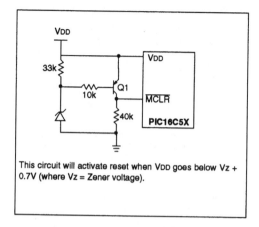

This circuit will activate reset when V$_{DD}$ goes below Vz + 0.7V (where Vz = Zener voltage).

Figure 2.6
Brown-out protection (I). (Reprinted with the permission of the copyright owner, Microchip Technology Incorporated © 1995. All rights reserved.)

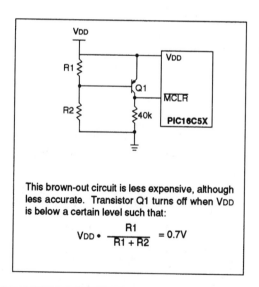

This brown-out circuit is less expensive, although less accurate. Transistor Q1 turns off when V$_{DD}$ is below a certain level such that:

$$V_{DD} \cdot \frac{R1}{R1 + R2} = 0.7V$$

Figure 2.7
Brown-out protection (II). (Reprinted with the permission of the copyright owner, Microchip Technology Incorporated © 1995. All rights reserved.)

2.2.3 I/O Ports

Using I/O pins, the micrcontroller contacts those devices that act in the real world. Such a part could be, for example, a key stroked by a user or a light emitting diode (LED) signalizing a certain status in a simple case.

The PIC16C56 has 12 I/O pins organized in PortA (RA3–RA0) and PortB (RB7–RB0). With the addition of PortC (RC7–RC0), the PIC16C57 has eight I/O pins more than the PIC16C56.

The equivalent circuit for an I/O port bit is shown in Figure 2.8. All port pins may be used for input or output operations and can be programmed individually. The data direction (that is, input or output) of each pin is controlled by I/O control registers. Register TRISA controls the data direction of PortA, Register TRISB controls the data direction of PortB, and so on. TRIS itself is an abbreviation for TRI-State. The TRIS registers are write-only. After reset, all I/O pins will be defined as input.

For input operation, the ports are non-latched. Each input operation reads the pin directly, independent of the defined data direction. The result of an output operation will be latched and remain unchanged until

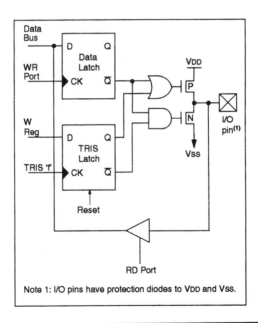

Note 1: I/O pins have protection diodes to VDD and Vss.

Figure 2.8
Equivalent circuit for a single I/O pin. (Reprinted with the permission of the copyright owner, Microchip Technology Incorporated © 1995. All rights reserved.)

the output latch is rewritten. To switch the output to the pin the corresponding bit of the TRIS register must be set to zero. To output a Lo level (0) at a certain pin the n-channel field-effect transistor (FET) is switched on, while to output a Hi level (1) the p-channel FET is switched on. If the corresponding I/O control-register bit is set, then both FET switches are in the high-impedance state.

Internal diodes are connected between the pin and V_{CC} and the pin and GND and protect the microcontroller circuit against electrostatic discharge and/or electrical overstress.

2.2.4 Internal Timer

The PIC16C56 and PIC16C57 contain a **Real-Time Clock/Counter** (RTCC) and a **Watch Dog Timer** (WDT). One 8-bit counter works as a prescaler for the RTCC or, alternatively, as a postscaler for the Watch Dog. In the product documentation, Microchip calls this counter a prescaler.

The Watch Dog Timer is a free-running on-chip RC oscillator and works without any additional external components. The WDT continues running when the clock has stopped. With the help of the WDT it is possible to generate a reset independent of whether the controller is working or sleeping.

The Watch Dog period normally amounts to 18 ms, depending on temperature, supply voltage, and process deviations. However, it is possible to generate periods up to a maximum of about 2.3 sec, by using the prescaler.

The RTCC register can be written and read by the program in the same way as any other register. In addition, its contents can be incremented by an external signal edge applied to the T0CKI pin or by the internal instruction cycle clock ($f_{OSC}/4$). The prescaler can divide both clock signals for incrementing the RTCC.

2.3 Technical Data of PIC16C56 and PIC16C57

This chapter summarizes some technical data of PIC16C56 and PIC16C57. However, that does not mean that Microchip's product documentation could be replaced by this information.

The pin configurations for PIC16C56 and PIC16C57 in SSOP (plastic surface) package are shown in Figure 2.9. The pin configurations for the other packages differ a little. A description of these pins is given in Table 2.3.

Table 2.4 shows the maximum ratings for some electrical parameters, and Table 2.5 shows some of the most important DC parameters.

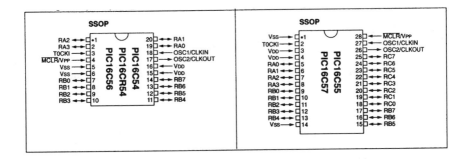

Figure 2.9
Pin configuration of PIC16C56 and PIC15C57 in SSOP package. (Reprinted with the permission of the copyright owner, Microchip Technology Incorporated © 1995. All rights reserved.)

Table 2.3
Pin description of PIC16C56 and PIC16C57

Pin	Function
RA3–RA0	I/O port A
RB7–RB0	I/O port B
RC7–RC0	I/O port C (PIC16C57 only)
T0CKI	Real-time clock/counter input
/MCLR	Master clear (reset)
OSC1	Oscillator input
OSC2/CLKOUT	Oscillator output
V_{DD}	Power supply
V_{SS}	Ground

Table 2.4
Maximum rating for PIC16C56 and PIC16C57

Characteristic	Maximum ratings
Voltage on any pin with respect to V_{SS} (except V_{DD} and /MCLR)	-0.6 V to $V_{DD} + 0.6$ V
Voltage on V_{DD} with respect to V_{SS}	0 to 9.5 V
Voltage on /MCLR with respect to V_{SS}	0 to 14 V
Max. output current sinked by any I/O pin	25 mA
Max. output current sourced by any I/O pin	20 mA
Max. output current sinked by a single I/O port	50 mA
Max. output current sourced by a single I/O port	40 mA

Table 2.5
DC characteristics for all pins except power supply

Characteristic	Symbol	Minimum	Type	Maximum	Unit	Conditions
Input lo voltage						
I/O pins	V_{IL}	V_{SS}	0.20 V_{DD}		V	Pin at tri-state
/MCLR pin	V_{ILMC}	V_{SS}	0.15 V_{DD}		V	
T0CKI pin	V_{ILRT}	V_{SS}	0.15 V_{DD}		V	
OSC1	V_{ILOSC}	V_{SS}	0.30 V_{DD}		V	
Input Hi voltage						
I/O pins	V_{IH}	0.45 V_{DD}		V_{DD}	V	For all V_{DD}
	V_{IH}	2.0		V_{DD}	V	$4.0 < V_{DD} < 5.5$ V
	V_{IH}	0.36 V_{DD}		V_{DD}	V	$V_{DD} > 5.5$ V
/MCLR pin	V_{IHMC}	0.85 V_{DD}		V_{DD}	V	
T0CKI pin	V_{IHRT}	0.85 V_{DD}		V_{DD}	V	
OSC1	V_{IHOSC}	0.70 V_{DD}		V_{DD}	V	
Input leakage current						For $V_{DD} < 5$ V
I/O pins	I_{IL}	-1	0.5	1	µA	$V_{SS} < V_{PIN} < V_{DD}$, Pin tri-state
/MCLR pin	I_{ILMCL}	-5			µA	$V_{PIN} = V_{SS} + 0.25$ V
/MCLR pin	I_{ILMCH}		0.5	5	µA	$V_{PIN} = V_{DD}$
T0CKI pin	I_{ILRT}	-3	0.5	3	µA	$V_{SS} < V_{PIN} < V_{DD}$
OSC1	I_{ILOSC1}	-3	0.5	3	µA	$V_{SS} < V_{PIN} < V_{DD}$
Output voltage						For $V_{DD} = 4.5$ V
I/O pin	V_{OL}			0.6	V	$I_{OL} = 8.7$ mA
	V_{OH}	$V_{DD} - 0.7$ V			V	$I_{OH} = -5.4$ mA

In the previous chapter, Microchip's PIC16C56 and PIC16C57 micro-controllers were described. The PIC16C56 is the base for the BASIC Stamp I and the PIC16C57 is the base for the BASIC Stamp II. This chapter describes the changes needed to get a BASIC Stamp from a PIC16C5x.

Parallax developed firmware and development systems for both BASIC Stamps. The firmware is burned in the OTP-EPROM (one-time programmable) of the PIC16C5x. With this firmware it is possible to interpret BASIC Tokens saved in an external EEPROM. With only two small devices the size of a stamp, BASIC Stamp builds a single-board computer programmable in PBASIC (Parallax-BASIC).

The development software works on IBM-PC or compatible without special requirements. The BASIC Tokens are the result of the compiled program. How to work with the development system and the PBASIC instruction set will be explained later.

3.1 BASIC Stamp I Device

The BASIC Stamp I device exists in two different packages and builds two different BASIC Stamp I single-board computers.

The first BASIC Stamp I had an 18-pin PDIP (plastic dual in-line) package mostly marked with the string PBASIC 1.x, with the character x describing the firmware version. The actual version is now called PBASIC 1.4. The firmware history is listed in the appendix. Figure 3.1 shows the pin description for the PDIP package.

The SSOP package has 20 pins. Because this package is too small, for labeling you will often find the original Microchip marking. Figure 3.2 shows the pin description for the SSOP package. The additional pins are used for V_{CC} and V_{SS}.

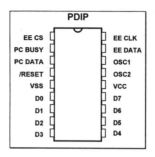

Figure 3.1
PDIP pinning BASIC Stamp I

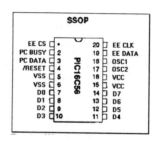

Figure 3.2
SSOP pinning BASIC Stamp I

At first look, the only difference between a PIC16C56 device and a BASIC Stamp I device seems to be different package marking information.

Port B provides the eight independent I/O pins. Port A is meant for EEPROM control and for data transfer between BASIC Stamp I and a PC. PC BUSY and PC DATA are two wires of a parallel PC printer port LPTx.

Three wires—EE CS (Chip Select), EE CLK (Clock for serial communication), and EE DATA (bi-directional data connection) build the interface between an external EEPROM and the microcontroller.

The interface to the PC is serial as well. The BUSY pin of the printer port serves for serial input and the data pin D0 (LSB) serves for serial output. This asynchronous serial interface between BASIC Stamp I and the PC works with a baud rate of 4800 bits/sec.

More interesting than the BASIC Stamp I device itself are the BASIC Stamp I single-board computer and the BASIC Stamp I modules. These modules already possess the required infrastructure for a complete micro-controller.

3.2 BASIC Stamp Version D

BASIC Stamp version D was the only Stamp for two years, and tens of thousands were sold. Surfing through the Internet shows that the original BASIC Stamp is used all over the world.

Figure 3.3 shows that the complete BASIC Stamp version D consists mainly of a printed circuit board with battery clips for a 9 V DC block (6F22), the BASIC Stamp I device and serial EEPROM in PDIP packages, a 3-pin I/O header as PC interface, a 14-pin I/O header as interface to the microcontroller, and a small prototyping area. After a download, the 256 byte EEPROM contains the tokens of the PBASIC application program and some fix data.

For the simple three-wire interface to the PC, a special cable is required. Parallax's programming package contains this cable but a homemade cable is good enough. Figure 3.4 shows the cable.

Figure 3.5 shows the schematic of BASIC Stamp version D. The V_{CC} and V_{SS} connections are not always shown.

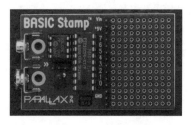

Figure 3.3
BASIC Stamp version D

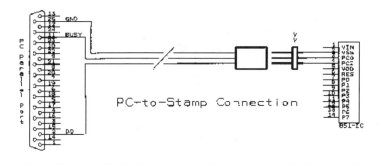

Figure 3.4
PC-BS1-Cable

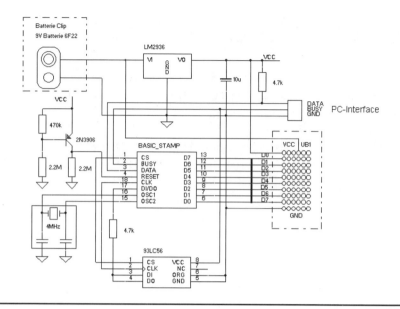

Figure 3.5
Schematic of BASIC Stamp version D

3.3 BS1-IC

BS1-IC is a module in a convenient 14-pin SIP (single in-line) package. The compactness of this module was achieved by use of surface mount devices (SMD). Figure 3.6 shows the entire BS1-IC module.

For module changes the compact BS1-IC has some advantages, but the handling of this small package is not so easy. For effective prototyping, a car-

Figure 3.6
BS1-IC

Figure 3.7
Carrier board with BS1-IC

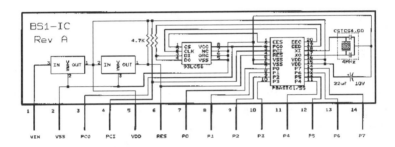

Figure 3.8
BS1-IC schematic

rier board for BS1-IC is needed. The carrier board including the BS1-IC is shown in Figure 3.7.

The circuitry of BS1-IC is almost the same as the circuitry of BASIC Stamp version D. A difference, however, is in the brown-out circuit. An integrated circuit provides a better defined reset. Figure 3.8 shows the schematic for BS1-IC.

3.4 BASIC Stamp II Device

The BASIC Stamp II device comes in a 28-pin SSOP package, as shown previously in Figure 2.9. Although this package is larger than the SSOP package

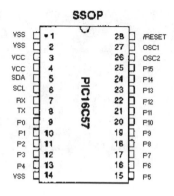

SSOP

```
        VSS  ⊏ •1      28 ⊐  /RESET
        VSS  ⊏  2      27 ⊐  OSC1
        VCC  ⊏  3      26 ⊐  OSC2
        VCC  ⊏  4      25 ⊐  P15
        SDA  ⊏  5   P  24 ⊐  P14
        SCL  ⊏  6   I  23 ⊐  P13
         RX  ⊏  7   C  22 ⊐  P12
         TX  ⊏  8   1  21 ⊐  P11
         P0  ⊏  9   6  20 ⊐  P10
         P1  ⊏ 10   C  19 ⊐  P9
         P2  ⊏ 11   5  18 ⊐  P8
         P3  ⊏ 12   7  17 ⊐  P7
         P4  ⊏ 13      16 ⊐  P6
        VSS  ⊏ 14      15 ⊐  P5
```

Figure 3.9
SSOP pinning BASIC Stamp II

for BS1, it also is too small for a new package marking. Therefore it often has the original Microchip marking. Figure 3.9 shows the pin description for the SSOP package.

Again, at first the external differences between a PIC16C57 device and a BASIC Stamp II device appear to be different package marking information.

Port B provides eight independent I/O pins and Port C an additional eight. Port A is meant for EEPROM control and data transfer between BASIC Stamp II and the PC. TX and RX are two wires of a serial PC port COMx. This asynchronous serial interface between BASIC Stamp II and the PC works with a baud rate of 9600 bits/sec.

Two wires—SCL (Clock for serial communication) and SDA (bi-directional data connection)—build the I²C interface between EEPROM and the microcontroller. The requirements for oscillator and reset circuitry are unchanged.

More interesting than the BASIC Stamp II device itself is the BASIC Stamp II module *BS2-IC*. This module already has the required infrastructure for a complete microcontroller.

3.5 BS2-IC

BS2-IC is a module in a convenient 28-pin DIL package. Again, using SMD achieves great compactness in this module. Figure 3.10 shows the entire BS2-IC module. Effective prototyping is supported by a carrier board for BS2-IC. The carrier board including the BS2-IC is shown in Figure 3.11.

Figure 3.10
BS2-IC

Figure 3.11
Carrier board with BS2-IC

The circuitry of BS2-IC consists of the same function blocks as the BS1 circuitry with only some components changing. The PIC16C57 microcontroller used here is programmed with the new *PBASIC2* firmware. After a download from a PC, a 2,048 byte EEPROM contains the tokens of the *PBASIC2* application program and some fix data. The used EEPROM has an I²C interface to the PIC16C57 microcontroller, and therefore needs only two wires for data exchange. A voltage regulator and a reset circuit give defined working conditions for the BS2-IC. If the voltage regulator is used, the supply voltage could be between 5 V and 15 V DC. If for any reason the regulated voltage drops under the level defined as the working range for the PIC16C57, the reset circuit will generate a stable power-down reset, preventing wild running.

Through the ATN input, a reset of the microcontroller can be generated exter-
nally. A defined start of the downloaded program is possible in this way. Figure
3.12 shows the whole circuitry of BS2-IC.

As mentioned in the last chapter, the connection between BS2-IC and the PC
is now a standard serial interface. Figure 3.13 shows the required connections.

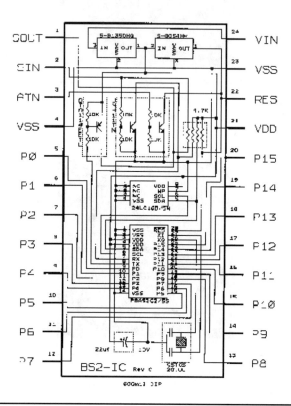

Figure 3.12
BS2-IC schematic

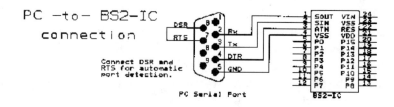

Figure 3.13
PC-BS2-Cable

Working with the carrier board for BS2-IC makes the connection quite easy. The carrier board contains a DSUB9 female connector, so a normal RS232 cable with DB9 on both sides may be used.

Figure 3.13 shows a little peculiarity. The handshake signals DSR and RTS of the RS232 Interface are connected. This helps the development system to find the COM port connected to BS2 equipment.

BASIC Stamp Development System   *4*

The BASIC Stamp development system includes all parts for programming and debugging BASIC Stamps. BASIC Stamp I and BASIC Stamp II have different systems, although the instruction sets contain many identical instructions.

4.1 System Requirements

To program BASIC Stamps, an IBM PC or compatible computer system with the following components is needed:

- 3.5-inch disk drive
- 128 kByte of RAM
- MS-DOS 2.0 or greater
- Parallel port, for BASIC Stamp I
- Serial port (RS232), for BASIC Stamp II

A 9-V battery 6F22 can be used to power the BASIC Stamps. Or a 5- to 15-V DC power supply may be used, but it is important to use the correct polarity. If there is a constant 5-V DC power supply, it can be connected directly to the +5-V pin of the BASIC Stamps. The voltage to this pin is limited to 6-V maximum. Otherwise, the circuitry may be damaged.

4.2 Parallax's Programming Package

The easiest way to get everything needed for BASIC Stamp programming is to purchase Parallax's programming package, which contains the following parts:

- BASIC Stamp I interface cable (parallel port DB25 to 3-pin header)
- BASIC Stamp I manual (including application notes)
- BASIC Stamp II interface cable (serial port DB9 to DB9)
- BASIC Stamp II manual
- 3.5-inch diskette

Both BASIC Stamps are supported by this programming package.

Parallax offers all PIC and BASIC Stamp related parts via the Internet too. Parallax's homepage can be found at

```
http//www.parallaxinc.com
```

At this URL, the whole BASIC Stamp offer can be found, including order information and a price list. Figure 4.1 shows the sales offer for the BASIC Stamp programming package.

For users looking for only one type of BASIC Stamp and/or for a less expensive way to get the parts listed above, Parallax offers the programming environment, manuals, and application notes for downloading via FTP at no cost. The cable for BASIC Stamp I must be homemade.

To provide an idea of the software parts of the programming package, Figure 4.2 shows a DIR listing of Parallax's disk containing all parts of the programming package.

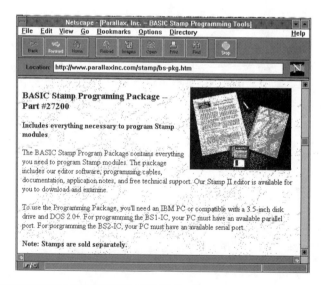

Figure 4.1
Parallax's BASIC Stamp programming package offered on the Internet

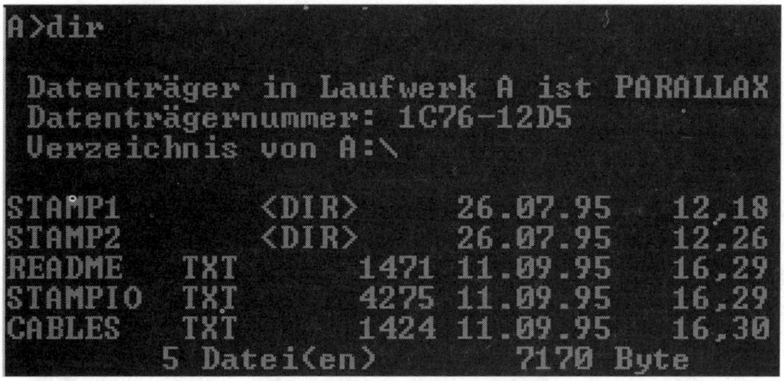

```
A>dir

 Datenträger in Laufwerk A ist PARALLAX
 Datenträgernummer: 1C76-12D5
 Verzeichnis von A:\

STAMP1          <DIR>      26.07.95    12,18
STAMP2          <DIR>      26.07.95    12,26
README    TXT       1471 11.09.95    16,29
STAMPIO   TXT       4275 11.09.95    16,29
CABLES    TXT       1424 11.09.95    16,30
          5 Datei(en)            7170 Byte
```

Figure 4.2
DIR listing of programming package

Each BASIC Stamp has its own directory, and some general information is given as text files.

4.3 Preparations for BASIC Stamp I Program Development

Before a first look at the development system, the BASIC Stamp I must be connected with the PC. All instructions provided here are specifically for the BS1-IC, but working with other BASIC Stamp I hardware should be no problem.
   To connect BASIC Stamp I to the PC, follow these steps:

• Plug the BS1-IC into the carrier board. When plugged into the carrier board, the component side of BS1-IC should face the battery clips.
• The interface cable for BASIC Stamp I has two ends, one with DB25 and the other with a simple 3-pin connector. Plug the remaining end of the interface cable onto the 3-pin header on the carrier board. The cable and the carrier board each have the marking <<. The marking on the cable and board should match up.
• Supply power to the carrier board, either by connecting a 9-V DC battery to the battery clips or by providing an external power source.

   With the BASIC Stamp I connected to the PC and powered, BS1 software can be started.
   Figure 4.3 shows the subdirectory STAMP1 where the file STAMP.EXE can be found. This file is the key to programming BASIC Stamp I.

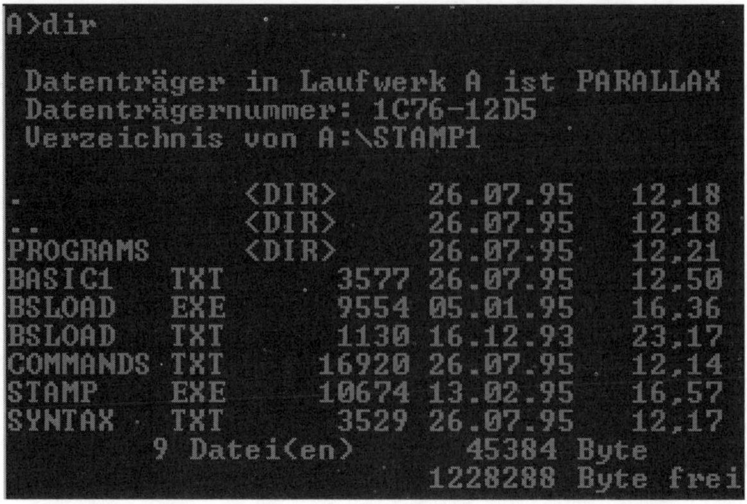

```
A>dir

Datenträger in Laufwerk A ist PARALLAX
Datenträgernummer: 1C76-12D5
Verzeichnis von A:\STAMP1

.               <DIR>        26.07.95    12,18
..              <DIR>        26.07.95    12,18
PROGRAMS        <DIR>        26.07.95    12,21
BASIC1   TXT       3577 26.07.95    12,50
BSLOAD   EXE       9554 05.01.95    16,36
BSLOAD   TXT       1130 16.12.93    23,17
COMMANDS TXT      16920 26.07.95    12,14
STAMP    EXE      10674 13.02.95    16,57
SYNTAX   TXT       3529 26.07.95    12,17
        9 Datei(en)              45384 Byte
                              1228288 Byte frei
```

Figure 4.3
Subdirectory STAMP1

From the DOS command line, start the program by typing the name of the program followed by the Enter key.

A:\STAMP1>STAMP<Enter>

The keystrokes are underlined.

After the program starts, a window opens for editing the PBASIC source code. Figure 4.4 shows this open window. In the top line are the commands for file handling (Load and Save), for compiling the source code and download-ing the tokens (Run), and for help and abort editing (Quit). The work with source code—editing and debugging—is described later.

4.4 Preparations for BASIC Stamp II Program Development

Preparation of the BASIC Stamp II equipment for program development is as easy as the preparation for BASIC Stamp I. All actions described here are related to the BS2-IC. Working with other BASIC Stamp II hardware is very similar.

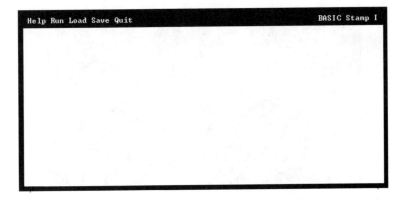

Figure 4.4
Edit window for BASIC Stamp I

To connect the BASIC Stamp II to the PC, follow these steps:

• Plug the BS2-IC into the carrier board. The BS2-IC plugs into a 24-pin DIP socket, located in the center of the carrier board. When the BS2-IC is plugged into the carrier board, the mark "Parallax BS2-IC" should be near the Reset button.
• The serial cable for connecting the BS2-IC to the PC must have DB9 connectors on both sides (male end for the BS2 and female end for the PC). The Parallax's programming package contains this required cable.
• Plug the female end of the serial cable into an available serial port on your PC.
• Plug the male end of the serial cable into the carrier board's serial port.
• Supply power to the carrier board, either by connecting a 9-V battery or by providing an external power supply.

With the BASIC Stamp II connected to the PC and powered, the BS2 software can be started.

Figure 4.5 shows the subdirectory STAMP2, where the file STAMP2.EXE can be found. This file is the key to programming the BASIC Stamp II.

From the DOS command line, start the program by typing the name of the program followed by the Enter key. The file EXAMPLE.BS2 is a very small test program, which can be loaded from the command line as follows.

```
A:\STAMP2>STAMP2 EXAMPLE <Enter>
```

The keystrokes are underlined.

BASIC Stamp Development System   37

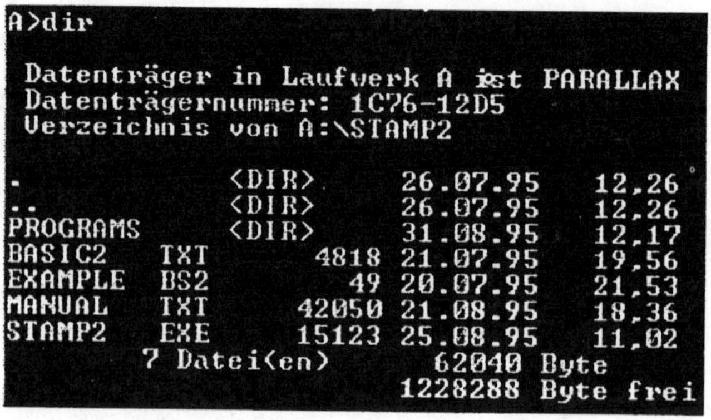

```
A>dir

Datenträger in Laufwerk A ist PARALLAX
Datenträgernummer: 1C76-12D5
Verzeichnis von A:\STAMP2

.              <DIR>      26.07.95    12.26
..             <DIR>      26.07.95    12.26
PROGRAMS       <DIR>      31.08.95    12.17
BASIC2   TXT       4818   21.07.95    19.56
EXAMPLE  BS2         49   20.07.95    21.53
MANUAL   TXT      42050   21.08.95    18.36
STAMP2   EXE      15123   25.08.95    11.02
       7 Datei(en)           62040 Byte
                           1228288 Byte frei
```

Figure 4.5
Subdirectory STAMP2

After the program starts, a window opens for editing the PBASIC source code of the loaded sample program. Figure 4.6 shows this open window.

4.5 Working with the Development System

For both BASIC Stamps, the required preparations for a program development were described earlier. This chapter describes editing and debugging of PBASIC programs with Parallax's development system and some differences programming both controllers.

4.5.1 Editing Programs

In Figures 4.4 and 4.6, the edit windows for both BASIC Stamp versions were shown. Each editor screen is dark blue, with one line across the top that gives needed hints. Except for the top line, the entire screen is available for entering and editing PBASIC programs.

Most functions of the common editor are easy to use. Using single keystrokes you can perform the following functions:

- Load, save, and run programs. Running a program means compiling the PBASIC source and downloading the generated tokens to the BASIC Stamp.
- Move the cursor by one character, one word, one line, one screen, or to the beginning or the end of the file.
- Highlight text in blocks of one character, one word, one line, one screen, or to the beginning or the end of the file.

Figure 4.6
Edit window for BASIC Stamp II - program EXAMPLE.BS2 loaded

- Cut, copy, and paste highlighted text.
- Search for and/or replace text.

Table 4.1 lists and explains the editor function keys.

Using Cut, Copy, and Paste   Cut, Copy, and Paste are effective methods to make changes in program text. The functional interface for these methods is the clipboard, an area of memory set aside by the editor. Cut and copy move the highlighted text to the clipboard; in addition, cut removes the text from its current location.

Text saved in the clipboard by cut or copy can later be pasted (inserted) at the cursor position anywhere in the program text.

Using Search and/or Replace   The editor has a function for searching and/or replacing with text. If you want to rename a variable, for example, it is not required to find each location of the name in the program text to make the change.

To set the search criteria, press **Alt-F** and enter the search string. For replace, enter the replace string, too. The replace string will be copied to the clipboard. If you only want to search, press Enter instead of entering the replacement string. During the search, you will have the option to replace individual occurences of the search string with the replacement string by pressing **Alt-V.** A further search function with the same search string can be started by pressing **Alt-N.**

Compile-time and Run-time Statements in PBASIC   Programming the BASIC Stamps means handling two categories of PBASIC statements: compile-time

Table 4.1

F1		Display editor help screen
Alt-H		Same as F1
Alt-R		Run program
Alt-L		Load program from disk
Alt-S		Save program to disk
Alt-M	BS2 only	Show memory usage maps
Alt-I	BS2 only	Show version number of PBASIC interpreter
Alt-P	BS1 only	Calibrate potentiometer scale
Alt-Q		Quit editor and return to DOS
Enter		Enter information and move cursor one line down
Tab		Same as Enter
Left arrow		Move cursor one character left
Right arrow		Move cursor one character right
Up arrow		Move cursor one line up
Down arrow		Move cursor one line down
Ctrl-Left		Move cursor left to the next word
Ctrl-Right		Move cursor right to the next word
Home		Move cursor to beginning of the current line
End		Move cursor to the end of the current line
Page Up		Move cursor up one screen
Page Down		Move cursor down one screen
Ctrl-Page Up		Move cursor to beginning of file
Ctrl-Page Down		Move cursor to end of file
Shift-Left		Highlight one character to the left
Shift-Right		Highlight one character to the right
Shift-Up		Highlight one line up
Shift-Down		Highlight one line down
Shift-Ctrl-Left		Highlight one word to the left
Shift-Ctrl-Right		Highlight one word to the right
Shift-Home		Highlight to beginning of line
Shift-End		Highlight to end of line
Shift-Page Up		Highlight one screen up
Shift-Page Down		Highlight one screen down
Shift-Ctrl-Page Up		Highlight to beginning of file
Shift-Ctrl-Page Down		Highlight to end of file
Shift-Insert		Highlight word at cursor
ESC		Cancel highlighted text
Backspace		Delete one character to the left
Delete		Delete character at cursor
Shift-Backspace		Delete from left character to the beginning of line
Shift-Delete		Delete to end of line
Ctrl-Backspace		Delete current line
Alt-X		Cut highlighted text and place in clipboard
Alt-C		Copy highlighted text to clipboard

Table 4.1 (continued)

Alt-V	Paste (insert) clipboard text at cursor
Alt-F	Find string (establish search information)
Alt-N	Find next occurence of string

statements and run-time statements. Compile-time statements are resolved during compilation of the program (after pressing **Alt-R** or **Alt-M**). No executable code will be generated. Run-time statements leave tokens (code) after compilation. These tokens will be executed by the BASIC Stamp itself at run-time.

The compile-time statements differ for the two BASIC Stamps. They will be described separately in the next two chapters. Compile-time statements arrange the access to memory via constants, variables, and create tables in EEPROM. Chapter 5 is reserved for the run-time statements—the PBASIC instruction set.

4.5.2 Programming BASIC Stamp I

BASIC Stamp I has 16 bytes of RAM devoted to I/O and the storage of variables. The arrangement of these memory cells and their names are shown in Table 4.2. *Do not forget:* These names are reserved words also.

Table 4.2
Organization of data memory

Word Name	Byte Name	Bit Name	Special Note
Port	Pins	Pin0 . . . Pin7	I/O pins; bit addressable
	Dirs	Dir0 . . . Dir7	I/O pin direction; bit addressable
W0	B0	Bit0 . . . Bit7	Bit addressable
	B1	Bit8 . . . Bit15	
W1	B2		
	B3		
W2	B4		
	B5		
W3	B6		
	B7		
W4	B8		
	B9		
W5	B10		
	B11		
W6	B12		Used by GOSUB instruction
	B13		

Port is a 16-bit word that is composed of two bytes, Pins and Dirs. Pins describes the I/O pins of BASIC Stamp I, and Dirs is the data direction (input or output) of each bit. A "0" in one of these bits causes the corresponding I/O pin to be an input. Otherwise a "1" causes the corresponding I/O pin to be an output.

Normally the I/O declaration happens at the beginning of a program. For example,

```
dirs = %00001111
```

defines Bit7 to Bit4 as input and Bit3 to Bit0 as output. Bit7 is the most significant bit MSB (most left position) and Bit0 is the least significant bit LSB (most right position).

After defining the data direction, you can read and write these pins. Some examples for reading and writing should explain the I/O access.

`B2 = pins`	Read all pins into the variable B2
`Bit0 = Pin7`	Read pin7 into bit variable Bit0
`if Pin3 = 1 then start`	Read pin3 and jumps to start if Pin3 was "1"
`Pins = %11000000`	Sets Bit7 and Bit6 to "1" and all other pins to "0"

To avoid cryptic programs, labels should be used to refer to constants and variables. For defining constants and labels the directive SYMBOL exists. A few examples describe the declaration of constants and variables.

`symbol switch = pin0`	Defines a bit variable switch with location pin0
`symbol flag = bit0`	Defines a bit variable flag with location bit0
`symbol count = b2`	Defines a byte variable with location b2
`symbol data = w3`	Defines a word variable with location w3

A simple program illustrates the use of constants and variables for BASIC Stamp I.

```
symbol start = 1  ' defines constant start with value 1
symbol end   = 10 ' defines constant end   with value 10

symbol count = b2 ' defines variable count with
                    location b2
```

```
loop: for count = start to end
      toggle 1    ' toggles pin1 ten times
      next
```

The variable count works as a counter in the for... next loop. Because its value goes from one to ten, a byte variable is big enough for this purpose.

The 256-byte EEPROM can be used for the program and data. It will be affected only by initial download (after pressing **Alt-R**) or run-time modifications. It survives power-down.

An EEPROM not used by the application program can be used for non-volatile data storage. The EEPROM is arranged as Figure 4.7 shows.

Starting at location $FE, the program tokens will be stored while downloaded. In location $FF the address of the first free byte in EEPROM is stored. If the program code occupies the memory locations $FE to $86, then the address $85 in location $FF can be found.

The BASIC Stamp I development system has no security mechanism to avoid overwriting of data and program area. With the program STMPSIZE, Jon Williams wrote a tool for inspecting the EEPROM use of BASIC Stamp I.

To store data in EEPROM before downloading the program, the directive EEPROM exists. It can be used in two ways.

```
EEPROM location, data, data, data, ...
EEPROM data, data, data, ...
```

In the first EEPROM directive, the starting location for the data area is given. In the second directive no starting location is given, so data will be written starting at the next available location.

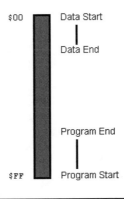

Figure 4.7
EEPROM of BASIC Stamp I

4.5.3 Programming BASIC Stamp II

BASIC Stamp II has 32 bytes of RAM devoted to I/O and the storage of variables. The arrangement of these memory cells and their names are shown in Table 4.3. *Do not forget:* These names are reserved words also.

INS, OUTS, and DIRS provide access to full 16-bit words for I/O. Access to byte, nibble, and bit for I/O operations is possible. IN(S) and OUT(S) describe the I/O pins of BASIC Stamp II, and DIR(S) the data direction (input or output) of each bit. A "0" in one of these bits causes the corresponding I/O pin to be an input. Otherwise a "1" causes the corresponding I/O pin to be an output.

Normally the I/O declaration happens at the beginning of a program. For example:

$$\text{dirs} = \%0000111111110000 \quad \text{or} \quad \begin{matrix} \text{dirh} = \%00001111 \\ \text{dirl} = \%11110000 \end{matrix}$$

Both declarations define Bit15 to Bit12 and Bit3 to Bit0 as input and Bit11 to Bit4 as output. Bit15 is the most significant bit MSB (most left position), and Bit0 is the least significant bit LSB (most right position).

Table 4.3
Data memory

Word	Word Name	Byte Name	Nibble Name	Bit Name	Special Note
$0	INS	INL INH	INA INB INC IND	IN0 ... IN15	pin input states; read-only
$1	OUTS	OUTL OUTH	OUTA OUTB OUTC OUTD	OUT0 ... OUT15	pin output latches; read/write
$2	DIRS	DIRL DIRH	DIRA DIRB DIRC DIRD	DIR0 ... DIR15	pin directions; read/write
$3					variable space; read/write
—					variable space; read/write
$F					variable space; read/write

After defining the data direction, you can read and write these pins. Some examples of reading and writing should explain the I/O access.

```
dw = pins              Read all pins into the variable DW (= Word)
carry = in7            Read pin7 into bit variable CARRY (= Bit)
if in3 = 1 then start  Read pin3 and jumps to start if Pin3 was "1"
out1 = %11000000       Sets Pin7 and Pin6 to "1" and Pin5 to Pin0
                          to "0"
```

An access via predefined names (IN_, OUT_ and DIR_) was possible only for the first three words. The access to the other locations in data memory must be through variables.

For BASIC Stamp II three compile-time statements are used for declaring variables (VAR), constants (CON), and EEPROM data (DATA).

The VAR statement assigns symbolic names to one (or more) RAM location(s). Variables could be declared at the beginning of a program as follows:

```
rhino var word    'makes "rhino" a word variable
dog var byte      'makes "dog" a byte variable
cat var nib       'makes "cat" a nibble variable
mouse var bit     'makes "mouse" a bit variable
```

but also

```
snake var bit(10)    'makes "snake" a 10-piece bit
                        variable
```

The compiler allocates memory for all words, bytes, nibbles, and bits in the unused RAM. By pressing **Alt-M** you can see a picture of the RAM allocation. Figure 4.8 shows this picture for the example above. The first three words are the reserved locations for I/O. Location W0 is reserved for the word variable "rhino." The byte variable "dog" is placed in the Hi byte of location W1. In the next nibble of W1 the nibble variable "cat" follows. The next bit (B3 in W1) is the bit variable "mouse" followed by ten bits of the variable "snake."

The arrangement of the declared variables in the memory is the cause of compilation. The user has no direct access to the memory cell itself.

Further features of the variable declaration are variables-in-variables and alias variables. An example describes these possibilities more simply than a long explanation.

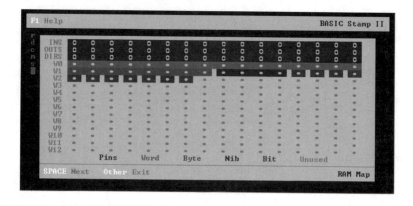

Figure 4.8
Memory map after declaration of some variables

Variables-in-variables:

```
adc var word              'declaration of word variable adc
sign var adc.highbit      'declaration of bit variable sign
                          'sign is the same as bit15 of adc
```

Alias variables:

```
time var word             'declaration of word variable time
delay_time var time       'alias variable delay_time
                          'alternate name but
                          'same memory location as time
```

There are simple building rules for variables:

```
symbol VAR size(array)
symbol VAR variable.modifier
```

Symbol, a unique name for the variable, will be defined later. Size is either WORD, BYTE, NIB, or BIT; it can be an expression, too. Array declares an array size. Variable is the name of an already defined variable, and modifier is used to build variable-in-variables. All variable modifiers are listed in Table 4.4.

Similar to the VAR statement, the CON statement defines constants and assigns to them symbolic names. The building rule is simple as for variables.

```
symbol CON expression
```

Table 4.4
Variable modifiers

Variable Modifier	Comment
LOWBYTE	Low byte of a word
HIGHBYTE	High byte of a word
BYTE0	Byte0 (low byte) of a word
BYTE1	Byte1 (high byte) of a word
LOWNIB	Low nibble of a word or byte
HIGHNIB	High nibble of a word or byte
NIB0	Nibble0 of a word or byte
NIB1	Nibble1 of a word or byte
NIB2	Nibble2 of a word
NIB3	Nibble3 of a word
LOWBIT	Low bit of a word, byte, or nibble
HIGHBIT	High bit of a word, byte, or nibble
BIT0	Bit0 of a word, byte, or nibble
BIT1	Bit1 of a word, byte, or nibble
BIT2	Bit2 of a word, byte, or nibble
BIT3	Bit3 of a word, byte, or nibble
BIT4	Bit4 of a word or byte
BIT5	Bit5 of a word or byte
BIT6	Bit6 of a word or byte
BIT7	Bit7 of a word or byte
BIT8	Bit8 of a word
BIT9	Bit9 of a word
BIT10	Bit10 of a word
BIT11	Bit11 of a word
BIT12	Bit12 of a word
BIT13	Bit13 of a word
BIT14	Bit14 of a word
BIT15	Bit15 of a word

Symbol is a unique name for the constant, and the expression gives a result calculated during compilation. Two examples illustrate the defining of constants:

```
level CON 10       'level is the same as the value 10
limit CON 10*4<<2  'limit will be calculated to 160
                   'during compilation
```

The expression after CON may contain the operators shown here in Table 4.5, and is resolved strictly left-to-right. The 2 kByte EEPROM of BASIC Stamp

Table 4.5
Operators for calculations
of constants

+	add
–	subtract
*	multiply
/	divide
<<	shift left
>>	shift right
&	logical AND
¦	logical OR
^	logical XOR

II can be used for program and data, too. It will be affected only by the initial download (after pressing **Alt-R**) or by run-time modifications. It survives power-down.

EEPROM not used by the application program can be used for non-volatile data storage. As Figure 4.9 shows, the EEPROM is arranged in this way: Starting at location $7FF the program tokens will be stored while downloaded. The BASIC Stamp II development system now has a security mechanism that avoids overwriting of data and program area. The compiler will give a message when you have a conflict.

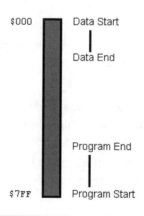

$000 — Data Start
Data End

Program End
$7FF — Program Start

Figure 4.9
EEPROM of BASIC Stamp II

The DATA statements are used to store data in unused EEPROM. Initially, the DATA location is set to 0. It is advanced by 1 for each byte declared. The building rules for EEPROM data are quite simple again.

```
symbol DATA data, data, data, ...
symbol DATA @pointer, data, data, data, ...
```

The optional symbol is a unique name for the data string following the DATA statement. The symbol name will be assigned to the start address of this data string and works as a data pointer. The DATA pointer may be altered by an @, followed by a new pointer.

Figure 4.10 shows the memory map of EEPROM for the following example of data definition:

```
DATA "Parallax"
DATA @$10,"BASIC Stamp II"
```

The first data string starts at location 0 and is eight characters long. The current data pointer would now go to location 8. The data pointer will be changed by the new data pointer to a value of $10. The start address of the second data string is therefore $10, too.

BASIC Stamp II has a further feature for defining EEPROM data. The data to this point were all defined data, i.e., they were fully declared and known at compilation time. For undefined data only memory allocation will be done, and values will not be assigned. Some examples show the differences. Figure 4.11 shows the memory map.

Figure 4.10
EEPROM after defining some data

Figure 4.11
EEPROM memory map with defined and undefined data

Defined data:

```
data "Parallax"
data 1,2,3,4,5
data word 1000
data $FF (8)
```

Undefined data:

```
data (16)
data word (8)
```

In the first line of EEPROM ($000 - $0FF) all entries seem correct. In the first half of the second line ($010 - $017), the eight bytes initialized to $FF follow. The rest of allocated memory should not be initialized because the next instructions describe undefined data. Nevertheless, the first eight bytes are initialized to 0. The reason for this behavior is the write mode for EEPROM. In Page Write Mode, 16 bytes can be programmed in one write cycle of a duration of 10 ms maximum.

A further interesting feature is the reference to tables shown in the next example:

```
string0 DATA "This is string0 . . . ",0
string1 DATA "This is string0 . . . ",0
string2 DATA "This is string2 . . . ",0
string3 DATA "This is string3 . . . ",0

strings DATA word string0, word string1,
            word string2, word string3
```

All strings are stored here as zero-terminated strings in EEPROM. Their data pointers are stored as an array following the address `strings`. An access to the strings for readout until character "0" occurs via address pointer `strings` and an offset.

4.6 Some Tools

The development system of BASIC Stamp I was not as complete as that now available for BASIC Stamp II. So it is no surprise that enthusiastic PBASIC programmers have designed some tools around the BASIC Stamp I. The programming tools described in this chapter are very useful in a BASIC Stamp I environment. All tools described here can be downloaded from Parallax's FTP server `ftp.parallaxinc.com`.

4.6.1 Program STMPSIZE

The section in this chapter called Programming BASIC Stamp I explained that the program area increases from high addresses to low addresses. The user memory, however, increases from low addresses to high addresses. To avoid overwriting, there must be a gap between these two memory areas.

Jon Williams wrote the program STMPSIZE to get a memory map from the 256 cells of external EEPROM for BASIC Stamp I. Figure 4.12 shows an example of a memory map of a file generated by the program STMPSIZE.

In this example, no user data are stored so we have used program memory in the map only. The top of the screen displays an overview about the usage of memory.

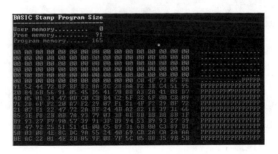

Figure 4.12
Memory map BASIC Stamp I

4.6.2 Program BSLOAD

The instruction set for BASIC Stamp I knows the instruction BSAVE (see Chapter 5 on the PBASIC Instruction Set). The result of compiling a source program *.BAS with the instruction BSAVE found at any location is a file CODE.OBJ.

This resulting file contains the program as tokens ready for interpreting by the token interpreter saved as firmware in the internal EPROM (OTP-ROM) of BASIC Stamp I during run-time. It is the 256-byte binary image of the Stamp's EEPROM.

The user should rename the file in DOS to some other appropriate name, such as RELAY.OBJ. Keep the OBJ extension, since the downloader assumes this automatically if nothing else is given. Once an *.OBJ file is made, it can be downloaded to a BASIC Stamp I via DOS by typing BSLOAD {filename}. The file will be downloaded to the BASIC Stamp I, and it runs as normal. Figure 4.13 shows the Help screen of BSLOAD.

This feature is of special interest for those programmers working on a custom project who do not want to publish their source program. The feature facilitates field upgrades to Stamp-based products. The customer needs a BASIC Stamp I cable, the BSLOAD program, and any *.OBJ files the customer wishes to supply.

4.6.3 Programs for Serial Communication

A main feature of the BASIC Stamps is the ease in handling serial communications. Serial communication here means communication from BASIC Stamp to BASIC Stamp as needed in BASIC Stamp networks, as well as communication from BASIC Stamp to an external serial interface, like that of a PC. In all cases, a tool is required to test these communication links. The shareware program PC2PC is very well-suited to test any implemented serial link.

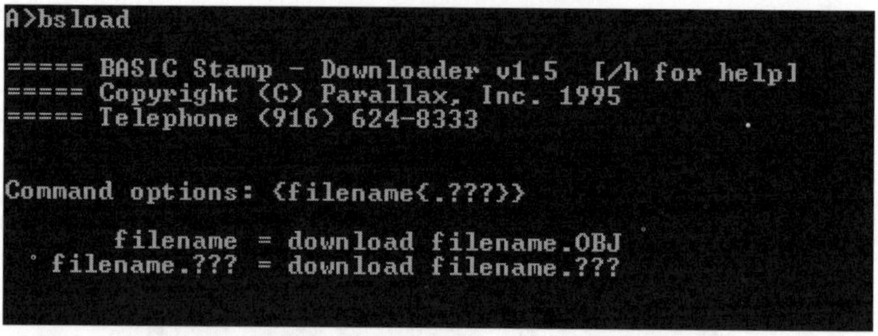

```
A>bsload

===== BASIC Stamp - Downloader v1.5  [/h for help]
===== Copyright (C) Parallax, Inc. 1995
===== Telephone (916) 624-8333

Command options: {filename{.???}}

        filename     = download filename.OBJ
    filename.???     = download filename.???
```

Figure 4.13
BASIC Stamp downloader

Figure 4.14 shows the opening screen of PC2PC and the possibilities for initialization. PC2PC works with COM1 or COM2 and with baud rates between 300 Baud and 9600 Baud. All other parameters are fixed. The serial mode specifies whether the program simulates transmitting only, receiving only, or bi-directional operation. In transmit operation, it is possible to echo the characters on the screen.

Another program for the same application is RS232MON. Figure 4.15 shows the working screen. Although the dialog is in German, handling of this program is very simple. For test purposes the TxD and RxD lines of COM1 are connected together. So you can see the transmitted characters in the transmit and the receive buffer too.

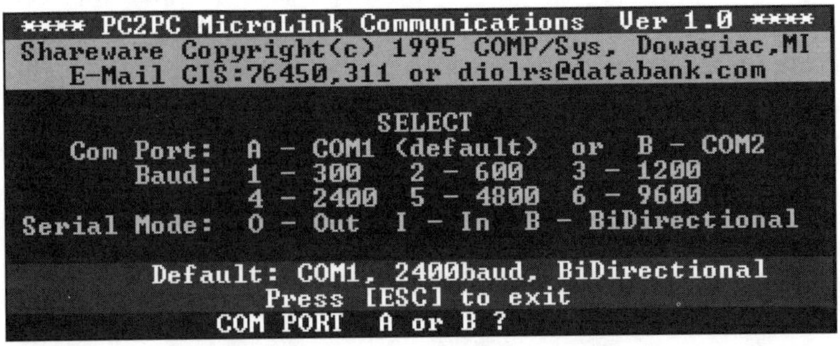

Figure 4.14
Shareware program PC2PC

Figure 4.15
Monitor program RS232MON

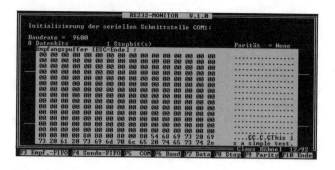

Figure 4.16
RS232MON - Inspecting the receive buffer

On the bottom line there are some function keys to display the receive or transmit buffer and for initializing the serial interface. Receive and transmit buffers are organized as ring buffers with a capacity of 256 bytes. Figure 4.16 displays a look at the receive buffer.

PBASIC for BASIC Stamp I and BASIC Stamp II

5

The PBASIC versions for the two BASIC Stamps differ only in a few instructions. Partly the reason is the different hardware; in other cases, changes were required for better handling of instructions. This chapter describes the PBASIC instruction set for both controllers. The differences will be indicated.

5.1 PBASIC Instruction Set

For a first overview of the PBASIC instruction set, Table 5.1 provides all instructions. Detailed explanations follow the table.

In the following pages, all PBASIC instructions will be described in detail. If there are any restrictions for use, that is indicated in the header.

The described procedure has two problems:

1. The notation of PBASIC for BASIC Stamp I and BASIC Stamp II differs slightly. All instructions for declaring variables and access to variables show these differences.
2. Unavoidably, in the explanation of some instructions, further instructions will be used that are themselves unexplained. In such situations the reader should simply look through this chapter for the necessary information. However, explanations and examples will be as simple and straightforward as possible.

BRANCH

```
BRANCH index,( label0, label1,...)
```

Function: Branch according to an index.

Table 5.1
PBASIC Instruction Set Overview

Instructions

B

ALL:	BRANCH *offset,(address,address...)*
BS1:	BSAVE
ALL:	BUTTON
	pin#,downstate,delay,rate,bytevar,targetstate,address

C

BS2:	COUNT *pin,period,variable*

D

BS1:	DEBUG *cls,"Text",cr,var,$var,%var,#var,#$var,#%var*
BS2:	DEBUG *CC, "Text", FORM VAR*
BS2:	DTMFOUT *pin,{ontime,offtime}*

E

ALL:	END

F

ALL:	FOR *var = start* TO *end* {STEP {-}*increment*}	NEXT {var}
BS2:	FREQOUT *pin#,milliseconds,freq1,{freq2}*

G

ALL:	GOSUB *label*	RETURN
ALL:	GOTO *address*

H

ALL:	HIGH *pin#*

I

ALL:	IF *variable ?? value*
	{AND/OR *variable ?? value ...*}
	THEN *label*
	;?? means = <> > < => <=
ALL:	INPUT *pin#*

L

ALL:	{LET} *variable = {-}value1 ?? value2 ...*
	*;?? means + - * / ** // MIN MAX &	^ &/ |/ ^/*
ALL:	LOOKDOWN *value,(value0,value1...),variable*
ALL:	LOOKUP *offset,(data,data...),variable*
ALL:	LOW *pin#*

N

ALL:	NAP *period*
ALL:	NEXT *{variable}*

O

ALL:	OUTPUT *pin#*

P

ALL:	PAUSE *milliseconds*
BS1:	POT *pin#,scale,variable*
ALL:	PULSIN *pin#,state,variable*
ALL:	PULSOUT *pin#,period*
ALL:	PWM *pin#,duty,cycles*

Table 5.1 (continued)

R

ALL: RANDOM *wordvariable*
BS2:   RCTIME *pin,state,variable*
ALL: READ *location,variable*
ALL: REVERSE *pin#*

S

BS1:   SERIN *pin#,baudmode,(qualifier,qualifier...)*
BS1:   SERIN *pin#,baudmode,(qualifier,...),{#}variable,...*
BS1:   SERIN *pin#,baudmode,{#}variable,{#}variable...*
BS1:   SEROUT *pin,baudmode,({#}data,{#}data...)*
BS2:   SERIN *pin#,baudmode,(qualifier,qualifier...)*
BS2:   SERIN *pin#,baudmode,(qualifier,...),{#}variable,...*
BS2:   SERIN *pin#,baudmode,{#}variable,{#}variable...*
BS2:   SEROUT *pin,baudmode,({#}data,{#}data...)*
BS2:   SHIFTIN *dpin,cpin,mode,[variable{\bits}...]*
BS2:   SHIFTOUT *dpin,cpin,[data{\bits}...]*
ALL: SLEEP *seconds*
BS1:   SOUND *pin#,(note,duration,note,duration...)*
BS2:   STOP

T

ALL: TOGGLE *pin#*

W

ALL: WRITE *location,data*

X

BS2:   XOUT *mpin,zpin,[house\keyorcommand{\cycles...}]*

Directives

BS1:   SYMBOL *name = variable*
BS1:   SYMBOL *name = constant*
BS2:   VAR *symbol VAR size(array)*
       symbol VAR variable.modifier
BS2:   CON *symbol CON expression*
BS2:   DATA *symbol DATA {@pointer,} data,data,data...*
BS1:   EEPROM *location,(data,data...)*
BS1:   EEPROM *(data,data...)*

It means:

| index | Pointer to a label of label list. |
| label | Element of a label list. |

Example:

```
symbol RxD  = 1 ' RxD at Pin1
symbol byte = b2
symbol value = b3
```

```
loop:   value = $FF ' initialize value to FFH
        ' Read serial interface
        serin RxD, N2400, byte
        ' Filter character "A", "B" and "C"
        lookdown byte, (65,66,67),value
        branch value, (aaa, bbb, ccc)
        debug cls
        goto loop

aaa:    debug cls,"Character A"   ' It was an A
        goto loop

bbb:    debug cls,"Character B"   ' It was a B
        goto loop

ccc:    debug cls,"Character C"   ' It was a C
        goto loop
```

Remark: Characters received from serial interface RxD will be searched for "A," "B," and/or "C." All other characters will be ignored. The resulting index (A = 0, B = 1, C = 2) controls the branching in the program.

BSAVE   **BS1 only**

BSAVE

Function: Stores the bit image of EEPROM (256 byte) as file CODE.OBJ.

Example:
```
        bsave   'Store the bit image as CODE.OBJ
        eeprom ("BSAVE-Test")
start:  toggle 0
        pause 500
        goto start
```

Remark: The file CODE.OBJ or its renamed copy is ready for downloading with program BSLOAD or for inspection with program STMPSIZE, both described already.

To avoid forgetting the relation to the application program, you should rename this program immediately. For example:

```
C:\>rename code.obj myprog.obj
```

BUTTON

```
BUTTON
   pin#,downstate,delay,rate,bytevar,targetstate,label
```

Function: Debounce button, auto-repeat, and branch if button is in the target state.

It means:

pin#	Pin number for input.
downstate	Element of a label list.
delay	Delay specifies the pressed time before auto-repeat (delay = 0: no debounce, no delay; delay = 255: only debounce, no repeat).
rate	Rate specifies the auto-repeat rate.
bytevar	Workspace for button. Must be cleared before first use of button.
targetstate	Targetstate specifies what state (0 = not pressed; 1 = pressed) the button should be in before a
label	branch can occur.

Example:

```
        symbol key = b2 'Defining workspace
        key = 0            'Initialization
loop:   button 1,0,9,1,key,0,wait
        low 0
        end
wait:   toggle 0
        pause 100
        goto loop
```

Remark: A lo-active key connected to pin1 will be polled. If the key is not pressed, a branch to label wait occurs. Pin0 toggles each 100 ms. If the key is pressed, Pin0 goes to Lo and the program ends. A delay of 10 × 100 ms and an auto-repeat rate of 1 × 100 ms will be generated.

COUNT   BS2 only

```
COUNT pin#,period,wordvar
```

Function: Count cycles on a pin for some milliseconds.

It means:

pin#	Pin number for count input.
period	Period specifies the count time in milliseconds.
wordvar	Variable to store result of counting.

Example:
```
count_input CON 1    'Count input is Pin1
count_time  CON 10   'Count time is 10ms
count_value VAR word
```
```
loop: count count_input,count_time,count_value
      debug dec? count_value
      goto loop
```

Remark: For a time of 10 ms Pin1 opens for counting. The counting result will be displayed via debug instructions on the PC's screen.

DEBUG

```
DEBUG  . . .
```

Function: Shows variables and messages for debugging purposes to the PC screen.

Attention: Different notations for BASIC Stamp I and BASIC Stamp II !

Example for BASIC Stamp I:
```
symbol value = b2
```

```
value = 123
debug cls,"DEBUG BS1",cr      'DEBUG BS1
debug value,cr                'VALUE = 123
debug "Value =",#value,cr     'Value = 123
debug #value,cr               '123
debug #$value,cr              '$7B
debug #%value,cr              '%01111011
```

Remark: The debug statements CLS clears the debugging screen and CR gives a carriage return. The comment in each line shows the resulting output on the debugging screen.

```
string VAR byte(4)
string(0)="A":string(1) = "B":string(2) = "C"
value VAR word:value = -123

debug cls,"DEBUG BS2",cr      'DEBUG BS2
debug rep "="\20,cr           '====================
debug str string,cr           'ABC
debug str string\2,cr         'AB
debug dec? value              'value = 65413
debug dec value,cr            '65413
debug dec3 value,cr           '413
debug sdec value,cr           '-123
debug sdec5 value,cr          '-00123
debug hex4 value,cr           'FF85
debug shex4 value,cr          '-007B
debug ihex4 value,cr          '$FF85
debug ishex4 value,cr         '-$007B
debug bin value,cr            '1111111110000101
debug sbin value,cr           '-1111011
debug ibin value,cr           '%1111111110000101
debug isbin value,cr          '-%1111011
```

Remark: There are more debug statements for BASIC Stamp II. CLS clears the debugging screen, HOME homes the cursor, BELL beeps the PC speaker, BKSP takes a backspace, TAB jumps to the next 8th column, and CR gives a carriage return. The comment in each line shows the resulting output on the debugging screen.

DTMFOUT   BS2 only

```
DTMFOUT pin#,{ontime,offtime,}[key,key,...]
```

Function: Output DTMF tones.

It means:

pin#	Pin number for output.
ontime	Default ontime is 200 ms.
offtime	Default offtime is 50 ms.
key	Telephone key (0 to 9 are the digits, 10 is *, 11 is #, and 12 to 15 are A-D (unavailable on normal telephones))

```
number data "4711"
key var nib(4)
i   var nib

    for i=0 to 3
        read number+i,key(i)
        'debug dec key(i),cr
    next

loop: dtmfout 0,250,100,
                    [key(0),key(1),key(2),key(3)]
      pause 2000
      goto loop
```

Remark: Four digits read from EEPROM will be dialed. Ontime and off-time are enhanced a little bit. Dialing is repeated for demonstration purposes. The output signal must be filtered by an RC circuit to achieve clean sine-waves. High-Z speakers may be driven with a coupling cap and a filter cap.

END

END

Function: End program.

Example:
```
low 0
pause 2000
high 0
end
```

Remark: Activates low-power mode and keeps I/O lines updated. About every 2.3 seconds the I/O lines will go to tri-state (High-Z) for about 18 ms. Approximately 50 µA average current will be consumed. END is terminated only by hardware reset.

FOR | NEXT

FOR *var* = *start* TO *end*   {STEP *increment*} | NEXT {*var*}

Function: Establish a FOR ... NEXT Loop.

It means:

var	Variable as loop counter.
start	Start value for counting.
end	End value for counting.
increment	Increment for loop counter. Default value is 1.

Example:
```
symbol i = b0

for i=0 to 8 step 2
    debug i
next
```

Remark: The increment value specifies a step value other then the default of 1. For BS1 the increment can be positive or negative. For BS2 only positive increments are allowed. To add or subtract the increment value, a to/from loop counter is determined at runtime. This permits going from 10 to 0 without specifying a negative increment value. FOR . . . NEXT loops can be nested up to 8 deep for BS1 and up to 16 deep for BS2.

FREQOUT   BS2 only

```
FREQOUT pin#,milliseconds,freq1,{freq2}
```

Function: Outputs one or two sine-waves for some time.

It means:

pin#	Pin number for tone output.
milliseconds	Determines the duration of tone output.
freq1	Specifies the first frequency in Hz (0...32768 Hz).
freq2	Is optional and specifies a second frequency.

Example:
```
melody data word 1000, word 2000,
            word 4000, word 2000

tone var word
tone_lo var tone.lowbyte
tone_hi var tone.highbyte
i    var nib
```

```
for i=0 to 3
    read 2*i + melody,tone_lo
    read 2*i + melody + 1, tone_hi
    tone = tone_hi * 256 + tone_lo
    debug dec? tone
    freqout 0,250,tone
next
```

Remark: The output signal may be filtered by an RC circuit to achieve a clean sine-wave(s). High-Z speakers may be driven with a coupling cap and a filter cap.

GOSUB | RETURN

```
GOSUB label | RETURN
```

Function: Go to a subroutine and return after running this subroutine.

It means:

label   Entry point of subroutine.

Example:
```
symbol i = b0

for i=0 to 3
    gosub show    'call subroutine show
next
end

show:  debug i      'subroutine for display
       return       'return to main loop
```

Remark: At the calling point the execution point is stored and then a branch to SHOW occurs. When a RETURN is encountered, execution continues at the instruction following the GOSUB. For BS1 there are 16 GOSUBs possible. For BS2, GOSUBs may be nested up to four deep and 255 in all.

GOTO

```
GOTO label
```

Function: Go to a new point in program.

It means:

label   Label for branch.

Example: `start: toggle 0`
                `        pause 1000`
                `        goto start`

Remark: On the last line of the program, a branch to `start` will occur.
               Execution continues at the next instruction.

HIGH

`HIGH pin#`

Function: Changes a pin to output and set to High.

It means:

pin#   Pin number for output High.

Example: `start: low 0`
                `        pause 200`
                `        high 0`
                `        pause 200`
                `        goto start`

Remark: Pin0 is set to High and Low, alternating.

IF | THEN

`IF variable ?? value {AND/OR variable ?? value ...}`
`THEN label`

Function: Branch conditionally.

It means:

variable   Variable with value for comparison.
??   Comparison operation.
value   Comparison value.
label   Label for branch.

Example: `start:input 5`   `'Read Pin5`

               `wait: if Pin5 = 1 then work 'Wait for Pin5 = 1`
                     `goto wait`

               `work: toggle 0             'Reaction`
                     `pause 1000`
                     `goto start`

Remark: If the result of the expression is true, execution will be continued at
               `label`. With logical operators some comparisons could be joined.

INPUT

`INPUT pin#`

Function: Changes a pin to input.

               It means:

               pin#   Pin number for input.

Example: `high 0`   `'Output Hi at Pin0`
               `input 0`   `'Change Pin0 as an input`
               `debug Pin0`   `'Read and display Pin0`
               `Pin0 = 0`   `'Clear variable Pin0`
               `debug Pin0`   `'Read and display Pin0`
               `reverse 0`   `'Change Pin0 to Output`
               `debug Pin0`   `'Read and display Pin0`
               `end`   `'End of program`

Remark: Pin0 is set from output to input.

LET

`{LET} var = value1 ?? value2 ?? ...`

Function: Stores the result of an operation to a variable. Valid operations
               are $+$, $-$, $*$, $/$, $**$, $//$, MIN, MAX, &, $|$, $\wedge$, &$/$, $|/$, $\wedge/$

               It means:

var	Result of operation
value1	First Operand
value2	Second Operand

Example:
```
symbol x = 10    'Constant
symbol y = b1    'Variable

y = 7 * x        'Arithmetic operation
```

Remark: All calculations will be done strictly from left to right. Later in this chapter, beginning on page 134, is a more detailed explanation.

LOCKDOWN

```
LOCKDOWN value,??[value0,value1, . . . ],var
```

Function: Lockdown a value and return an index.

For BS1 only = is a valid comparison operator and will be not written. For BS2 =, <>, >, <, <=, => are valid.

It means:

value	Value for comparison.
??	Comparison operator. Default is =.
value0	Comparison value0.
value1	Comparison value1.
var	Variable contains index from comparison.

Example:
```
symbol RxD = 1 ' RxD at Pin1
symbol byte = b2
symbol value = b3

loop: value = $FF    ' initialize value to FFH
      ' Read serial interface
      serin RxD, N2400, byte
      ' Filter character "A", "B" and "C"
      lookdown byte,(65,66,67),value
      branch value, (aaa, bbb, ccc)
      debug cls
      goto loop
```

```
aaa:    debug cls,"Character A"   ' It was an A
        goto loop

bbb:    debug cls,"Character B"   ' It was a B
        goto loop

ccc:    debug cls,"Character C"   ' It was a C
        goto loop
```

Remark: A comparison is made between value and value0; if the result is true, the index 0 is written into var. If the comparison was false, a further comparison is made with value1. If this comparison is true the index 1 is written into var, and so on. If no result is true the variable var is unaffected.

LOCKUP

```
LOOKUP index,[value0,value1,...],var
```

Function: Look up a variable according to an index.

It means:

index	Value position in the value list.
value0	Value at index 0.
value1	Value at index 1.
var	Result of lookup operation.

Example:
```
symbol i     = b0
symbol value = b1

for i=0 to 7
    lookup i,(1,2,4,8,16,32,64,128),value
    debug i,value
next
```

Remark: If index = 0 value0 will be written into var. If index = 1 value1 will be written into var. If index exceeds the number of value entries, then var will not be affected.

LOW

```
LOW pin#
```

Function: Changes a pin to output and sets it Low.

It means:

pin#   Pin number for output Low.

Example: start: low 0
                   pause 200
                   high 0
                   pause 200
                   goto start

Remark: Pin0 is set to High and Low alternating.

NAP

NAP period

Function: Nap for a short period.

It means:

period   Factor for duration of nap (period=0 to 7).

p	0	1	2	3	4	5	6	7
T	0.018	0.036	0.072	0.140	0.290	0.580	1.200	2.300

Example: symbol long = 7

loop: nap long   'Nap about 2.3 sec
        goto loop

Remark: Enter low-power mode for a short period. When the period is over, the pins will go to tri-state (High-Z) for about 18 ms and execution will continue at the next instruction.

OUTPUT

OUTPUT pin#

Function: Changes a pin to output.

It means:

pin#   Pin number for output.

Example:
```
high0       'Output Hi at Pin0
input 0     'Change Pin0 as an input
debug Pin0  'Read and display Pin0
Pin0 = 0    'Clear variable Pin0
debug Pin0  'Read and display Pin0
output 0    'Change Pin0 to Output
debug Pin0  'Read and display Pin0
end         'End of program
```

Remark: Pin0 is set from input to output.

PAUSE

```
PAUSE period
```

Function: Pause for some ms.

It means:

period   Factor for duration of pause (period = 0 . . . 65535).

Example:
```
start: low 0
       pause 200      'Pause for 200 ms
       high 0
       pause 200      'Pause for 200 ms
       goto start
```

Remark: A delay of 200 ms occurs. The accuracy is dependent on the oscillator frequency. The function call adds additional time to the delay from pause.

POT   **BS1 only**

```
POT pin#, scale, var
```

Function: Reads a potentiometer with a value between 5 kΩ and 50 kΩ.

It means:

pin#   Pin number for connecting RC circuit.
scale   Factor for scaling measurement range.
var   Result of resistor measurement.

Example:

```
symbol value = b1     'Value of resistor
   symbol scale = 126 'Scaling factor
   symbol limit = 128 'Limit
   symbol lamp  = 0   'Lamp at Pin0
   symbol sensor= 1   'Sensor at Pin1

loop: pot sensor,scale,value 'Measure resistor
      if value < limit then lamp_off

lamp_on:
      high lamp            'Lamp On
      goto loop
lamp_off:
      low lamp             'Lamp Off
      goto loop
```

Remark: A resistor value in a range from 5 kΩ to about 50 kΩ can be measured. The resistor value will be calculated from measurement of
the time constant of an RC circuit.

One side of the resistor is connected to the pin and the other side
is connected through a cap (0.1 μF) to the ground.

For finding the best scale value, the editor of BS1 has a calibration option. After connecting the resistor to be used with the
POT instruction, press **Alt-P.** A calibration window opens and asks
for the pin used. The editor downloads a small calibration program (overwriting the application). In the next window you will
find two numbers indicating scale and value. Now adjust the resistor (or potentiometer) until the smallest possible number is shown
for scale. That's all.

To verify the scale factor found by calibration, press the spacebar.
The scale factor will be fixed and the resistor changes should give
values between 0 and 255.

This procedure can be repeated until the best scale value is found.

PULSIN

```
PULSIN pin#,state,var
```

Function: Measures the width of an incoming pulse.

It means:

pin#	Pin number for pulse input.
state	Measured pulse state.
var	Result of measurement (byte or word variable).

Example:

```
symbol puls_in = 2   'Pulse input Pin2
symbol state   = 1   'Trigger with 0-1
symbol value   = w1

loop: pulsin puls_in,state,value
      debug value
      goto loop
```

Remark: The pin will be placed in input mode and and the edge of a pulse, defined in state, will be awaited and measured. For BS1 the resolution is 10 µs and for BS2 it is 2 µs. A 16-bit internal timer is used. For BS1 the maximum pulse width measured with `pulsin` is 655.35 ms for a word variable, or 2.55 ms for a byte variable. For BS2 the oscillator frequency is five times that of BS1, so the BS2 pulse widths must be divided by five. The waiting time for a start trigger is defined in the same way. BS1 waits 655.35 ms for a trigger in maximum, while BS2 waits a maximum of 131.07 ms.

PULSOUT

```
PULSOUT pin#, period
```

Function: Output a timed pulse by inverting a pin for a specified amount of time.

It means:

pin#	Pin number for pulse output.
period	Factor for duration of pulse (period = 0 . . . 65535)

Example:

```
symbol pulse = 0        'Pulse output Pin0
symbol time  =  10000   'Duration 10000*10µs

low pulse               'Pulse=Low
pulsout pulse,time      'Pulse=High for 100 ms
```

Remark: The state will be switched to the opposite for a specified time. For BS1 the resolution is 10 µs and for BS2 it is 2 µs. A 16-bit internal timer is used. For BS1 the maximum pulse width for `pulseout` is 655.35 ms. For BS2 the oscillator frequency is five times that of BS1, so the pulse width for BS2 must be divided by five.

PWM

```
PWM pin#,duty,cycles
```

Function: Outputs a pulse-width modulated signal for some time.

It means:

pin#	Pin number for output.
duty	Specifies the analogue voltage (0-255 = 0-5V)
cycles	Number of cycles for output (0-255)

Example:

```
            symbol time = 5000   'Waiting time

loop: pwm 1,51,100    'U = 1V
      pause time      'Wait 5 sec.
      pwm 1,102,100   'U = 2V
      pause time      'Wait 5 sec.
      pwm 1,153,100   'U = 3V
      pause time      'Wait 5 sec.
      pwm 1,204,100   'U = 4V
      pause time      'Wait 5 sec.
      goto loop       'Repeat endless
```

Remark: PWM outputs pulse-width modulation on a pin for a defined number of cycles. For BS1 one cycle measures about 5 ms, for BS2 it measures about 1 ms. After the specified number of cycles the pin will go to input state again.

An analog voltage can be measured if a resistor is connected to the pin and the other side of the resistor goes through a cap to the ground. The analog voltage can be measured over the cap. Since the cap gradually discharges, PWM should be executed periodically to update and/or refresh the analog voltage.

RANDOM

```
RANDOM wordvar
```

Function: Generates a pseudo-random number.

It means:

wordvar   Wordvariable

Example:
```
symbol value = w1 'Wordvariable for random
symbol i = b4

value = 999          'Initialize value to 999
for i=0 to 9
     random value
     debug value
next
```

Remark: A pseudo-random number is generated, dependent on the initial value. For random initialization an RTC can be read, for example.

RCTIME   **BS2 only**

```
RCTIME pin#,state,wordvar
```

Function: Measures an RC charge/discharge time.

It means:

pin#   Pin number for input.
state   Input state
var   Result of measurement (Word variable)

Example:
```
pot var word

loop: high 0
      pause 1
      rctime 0,1,pot
      degub ? pot
      goto loop
```

Remark: To convert the value of a resistor to a digital number, one end of the resistor goes to the ground and the other through a cap to VCC. Between the connection of resistor and capacitor and the pin, a small resistor should be connected. This resistor prevents a possible short between a high pin and the ground (if the measured resistor goes to zero).

In the program above, Pin0 will be changed to High for 1 ms followed from rctime. The voltage at Pin0 falls from 5 V to the ground. If the voltage reaches a level of about 1.4 V, the input state switches from Hi to Lo and counting will be terminated. The counting result will be stored in the word variable. Time will be measured in steps to 2 μs.

READ

```
READ loc,var
```

Function: Reads a byte from EEPROM.

It means:

loc	Memory location in EEPROM (BS1: 0-255; BS2: 0-2047)
var	Variable for read byte value

Example:
```
symbol last_token = b2
symbol max_data   = b3

read 255,b2 'Read address of last token for BS1
max_data = last_token -1
debug last_token, max_data
```

Remark: For BS1, the address of the last token is saved. Knowing this value can prevent the overriding of EEPROM data.

REVERSE

```
REVERSE pin#
```

Function: Flips the pin direction.

It means:

pin#	Pin number input or output.

Example:
```
high 0      'Output Hi at Pin0
input 0     'Change Pin0 as an input
debug Pin0  'Read and display Pin0
Pin0 = 0    'Clear variable Pin0
debug Pin0  'Read and display Pin0
reverse 0   'Change Pin0 to Output
debug Pin0  'Read and display Pin0
end         'End of program
```

Remark: Pin0 is set from input to output.

SERIN   **BS1 only**

```
SERIN pin#,baudmode,(qual,qual,...)
SERIN pin#,baudmode,{#}var,{#}var,...
SERIN pin#,baudmode,(qual,qual,...),{#}var,{#}var
```

Function: Initialize serial input and then wait for optional qualifiers and/or variables.

It means:

pin#	Pin number for serial input.
baudmode	Parameters for serial communication. Fix parameters are: 8 Databits, 1 Startbit, no parity. Changeable parameters see table under remark.
qual	Qualifiers are optional characters which must be received before execution can continue.

var		Variables are byte(s) to store the received characters. If a # precedes a variable name the serin will convert numeric text into a value to fill the variable (1 2 3 —> [7B] —> {)		

Example:

```
symbol RxD   = 1           ' RxD at Pin1
symbol baudmode = N2400 '2400 Baud invert
symbol rec_char = b0      'Data byte

loop: serin RxD, baudmode, ("CD"), rec_char
      debug rec_char
      goto loop
```

Remark: Stops the program until the characters CD are received serially through Pin1 with inverted polarity and 2400 baud. Once both qualifiers have been received, the next character will be stored in variable rec_char for later display via debug.

Baudmode (use # or symbol in serin):

#	Symbol	Baudrate	Signal Level
0	T2400	2400	true
1	T1200	1200	true
2	T600	600	true
3	T300	300	true
4	N2400	2400	inverted
5	N1200	1200	inverted
6	N600	600	inverted
7	N300	300	inverted

SERIN   **BS2 only**

```
SERIN rpin#{\fpin#},baudmode,
      {plabel,}{timeout,tlabel,}[inputdata]
```

Function: Input data serially.

It means:

rpin#	Pin number for serial input. 0-15 are I/O pins, 16 is internal RxD.

fpin#	Pin number for flow control.
baudmode	Parameters for serial communication. Description see `serout` for BS2.
plabel	Branch to `plabel` if a parity error occurs. Parity mode must be enabled.
timeout	`Timeout` in milliseconds, specifies how long to wait for receiving character until branch to `tlabel`.
tlabel	Branch to `tlabel` if timeout occurs.
inputdata	See conventions for inputdata under remark.

Example:

```
RxD con  0          'RxD at Pin0
baud con 396+$4000  'N2400
rec_char var byte

loop: serin RxD,baud,50,timeout,[rec_char]
      debug "Rec. char.:", HEX2 rec_char,CR
      goto loop

timeout:debug "TimeOut",CR
        goto loop
```

Remark: Conventions for input data:

```
variable
STR bytearray\L{\E}
SKIP L
WAITSTR bytearray
WAITSTR bytearray\L
WAIT (value, value,...)
DEC variable
SDEC variable
HEX variable
SHEX variable
IHEX variable
ISHEX variable
BIN variable
SBIN variable
IBIN variable
ISBIN variable
```

Note: If `rpin=16` (internal RxD), baudmode's bit15 and 14 are ignored. However, they will affect the operation of `fpin`.

SEROUT

BS1 only

```
SEROUT pin#,baudmode,({#}data,{#}data,...)
```

Function: Initialize a serial output port and transmit data.

It means:

pin#	Pin number for serial output.
baudmode	Parameters for serial communication. Fixed parameters are: 8 Databits, 1 Startbit, no parity. Changeable parameters see table under remark.
data	Data are byte(s) for output serial. If a # precedes a data name the serout will convert the data byte into numeric text up to five characters long ({--> 1 2 3). Otherwise a single byte will be transferred.

Example:
```
symbol TxD = 2      'TxD at Pin2
symbol baud = N2400 '2400 Baud, inverted
symbol tra_char = 7 'Data byte

loop:
serout TxD,baud,("Hello world",tra_char,13,10)
        pause 1000
        goto loop
```

Remark: The example above shows the serial output of a string limited by characters or quotation marks and a data byte followed by a CR/LF.

Baudmode (use # or symbol in serout):

#	Symbol	Baudrate	Signal Level
0	T2400	2400	true always driven
1	T1200	1200	true always driven
2	T600	600	true always driven
3	T300	300	true always driven
4	N2400	2400	inverted always driven
5	N1200	1200	inverted always driven
6	N600	600	inverted always driven
7	N300	300	inverted always driven
8	OT2400	2400	true open drain
9	OT1200	1200	true open drain

PBASIC for BASIC Stamp I and BASIC Stamp II   79

#	Symbol	Baudrate	Signal Level
10	OT600	600	true open drain
11	OT300	300	true open drain
12	ON2400	2400	inverted open source
13	ON1200	1200	inverted open source
14	ON600	600	inverted open source
15	ON300	300	inverted open source

SEROUT   BS2 only

```
SEROUT tpin#,baudmode,{pace,}[outputdata]
SEROUT
  tpin#\fpin#,baudmode,{timeout,tlabel,}[outputdata]
```

Function: Output data serially.

It means:

tpin#	Pin number for serial output. 0-15 are I/O pins, 16 is internal TxD.
fpin#	Pin number for flow control.
baudmode	Parameters for serial communication. See description under Remark.
pace	If no pin is defined for flow control, an optional pace in milliseconds may be declared to pace characters by this time.
timeout	Timeout in milliseconds, specifies how long to wait for receiving character until branch to tlabel.
tlabel	Branch to tlabel if timeout occurs.
outputdata	Conventions follow the debug instructions.

Example:
```
TxD   con 1            'TxD at Pin1
baud  con 396+$4000    'N2400
pace  con 100          'Pace = 100 ms

loop: serout RxD,baud,pace,["Hello world"]
      goto loop
```

Remark: A pace of 100 ms will be placed between all the characters of the transferred string.

Baudmode calculation:

$$\text{Bit Period} = int\left(\frac{1,000,000}{\text{Baud Rate}}\right) - 20$$

```
15 14 13 12 11 10  9  8  7  6  5  4  3  2  1  0
 |  |  |  |---------- Bit period --------------|
 |  |  |
 |  |  |
 |  |  0: 8 Data bits no parity
 |  |  1: 7 Data bits parity (add $2000)
 |  0: true polarity
 |  1: inverted polarity (add $4000)
 0: always driven
 1: open drain/source (add $8000; serout only)
```

SHIFTIN   **BS2 only**

SHIFTIN dpin#,cpin#,mode,[var{\bits},...]

Function: Shift bits in synchronously.

It means:

dpin#	Pin number for data input.
cpin#	Pin number for clock output.
mode	Operation mode (look at Remark).
var	Variable receives the data shifted-in.
bits	Optional number of bits. Default is eight.

Example:
```
ADres var byte    'A-to-D result: one byte.
CS    con  0      'Chip select is pin 0.
AData con  1      'ADC data output is pin 1.
CLK   con  2      'Clock is pin 2.

high CS           'Deselect ADC to start.

again:
  low CS                    'Activate the ADC0831.
  shiftin AData,CLK,msbpost,[ADres\9]
  high CS                   'Deactivate '0831.
  debug ? ADres             'Show us the result.
  pause 1000                'Wait a second.
goto again                  'Do it again.
```

Remark: Mode characterizes the direction for shifting and the position for the clock pulse. There are predefined words for the different modes. Mode is 0 (MSBPRE) for msb-first/pre-clock, 1 (LSBPRE) for lsb-first/pre-clock, 2 (MSBPOST) for msb-first/post-clock and 3 (LSBPOST) for lsb-first/post-clock.

SHIFTOUT   BS2 only

```
SHIFTOUT dpin#,cpin#,mode,[var{\bits},...]
```

Function: Shift bits out synchronously.

It means:

dpin#	Pin number for data output.
cpin#	Pin number for clock output.
mode	Operation mode (look at remark).
var	Variable to be shifted out.
bits	Optional number of bits. Default is eight.

Example:
```
DataP   con 0      'Data pin to 74HC595.
Clock   con 1      'Shift clock to '595.
Latch   con 2      'Moves data from shift
                   'register to output latch.
counter var byte 'Counter for demo program.

Again:
  Shiftout DataP,Clock,msbfirst,[counter]
  pulsout Latch,1        'Transfer to outputs.
  pause 50               'Wait briefly.
  counter = counter+1    'Increment counter.
goto Again               'Do it again.
```

Remark: Mode characterizes the direction for shifting. There are predefined words for the different modes. Mode is 0 (LSBFIRST) for lsb-first and 1 (MSBFIRST) for msb-first.

SLEEP

```
SLEEP seconds
```

Function: Enter Sleep Mode for a specified number of seconds.

It means:

seconds   Duration of sleep in Seconds (word var 0-65535)

Example:
```
symbol time = 10 'Sleep time about 10 s
high 0
sleep time
low 0
end
```

Remark: Enters low-power mode and keeps I/Os updated. Every ~2.3 s, the I/Os will be tristated for about 18 ms.

SOUND   BS1 only

```
SOUND pin#,(note0, duration0, note1,duration1,...)
```

Function: Generates square-wave notes with a given duration.

It means:

pin#   Pin number for tone output
note   Variable that specifies type and frequency (note = 0: silence; note = 1-127: proportional frequency; note > 127: white noise).
duration   Duration in unit of 12 ms

Example:
```
symbol tone = b1

start:tone = 1
loop: sound 1,(tone,5)
      tone = tone + 10
      if tone > 127 then start
      goto loop
```

Remark: The notes can vary in frequency from 94.8 Hz to 10,550 Hz.

STOP   BS2 only

```
STOP
```

Function: Stop execution.

Example:

```
            symbol tone = b1

   start:tone = 55
            sound 1,(tone,5)
            stop
```

Remark: Execution is frozen, but low-power mode is not entered. This is like end, except that the pins never go to tri-state (High-Z); they remain driven.

TOGGLE

```
TOGGLE pin#
```

Function: Changes a pin to output and toggle state.

It means:

pin#   Pin number for output.

Example:
```
start: low 0
loop:  pause 200
       toggle 0
       goto loop
```

Remark: State of Pin0 alternates permanently.

WRITE

```
WRITE loc,var
```

Function: Writes a data byte into EEPROM.

It means:

loc   Memory location in EEPROM
             (BS1: 0-255; BS2: 0-2047)
var   Variable for write byte value

Example:
```
symbol addr = 0
symbol value = b2
```

```
write addr,$AA
value = 0
read  addr,value
debug value

write addr,$55
value = 0
read  addr,value
debug value

end
```

XOUT   **BS2 only**

```
XOUT mpin#,zpin#,[house\keyorcommand{\cycles . . . }]
```

Function: Output X-10 powerline control codes to a PL513 or TW523 powerline interface module.

It means:

mpin#	Pin number for modulation control.
zpin#	Pin number for zero-crossing detection.
house	House code (0-15 is "A"-"P")
keyorcommand	Key number (0-15 is "1"-"16") or command
cycles	Optional number, use only for DIM or BRIGHT commands. Default is 2.

Example:
```
zPin con  0    ' Zero-crossing-detect
mPin con  1    ' Modulation-control pin

houseA    con 0    ' House code: 0=A, 1=B . . .
Unit1     con 0    ' Unit code: 0=1, 1=2 . . .
Unit2     con 1    ' Unit code: 1=2.

xout mPin,zPin,[houseA\Unit1]    'Talk to Unit1.
xout mPin,zPin,[houseA\uniton] 'Turn ON.
pause 1000                      'Wait a second.
xout mPin,zPin,[houseA\unitoff]'Turn OFF.
```

```
xout mPin,zPin,[houseA\Unit2]   'Talk to Unit2.
xout mPin,zPin,[houseA\unitoff]'Turn OFF.
xout mPin,zPin,[houseA\dim\10]  'Dim unit.
```

Remark:

X-10 Commands	Value	Powerline Interface Pin	BS2 Pin
===================		==================	
UNITON	%10010		
UNITOFF	%11010	==================	
UNITSOFF	%11100	1	zPin
LIGHTSON	%10100	2	GND
DIM	%11110	3	GND
BRIGHT	%10110	4	mPin

5.2 Explanations of Some PBASIC Instructions

Not all PBASIC instructions are quite as simple in their handling as they seem at first sight. Therefore, some instructions are explained in more detail in this chapter.

5.2.1 Query a Key with BUTTON

The BUTTON instruction has many parameters, and discussions with Parallax make clear that some BASIC Stamp users have trouble with this instruction. Jon Williams has developed a very clear explanation of the BUTTON instruction and has downloaded this as file BUTTON.ZIP to Parallax's BBS and FTP sites. The most important parts of his explanation are presented here.

Only a simple program and a bit of explanation are necessary to master the auto-repeat feature of this command.

Let's review the parameters of the BUTTON command:

```
BUTTON pin,downstate,delay,rate,bytevariable,targetstate,address
```

- **Pin** is a variable/constant (0-7) which specifies the I/O pin to use.
- **Downstate** is a variable/constant (0 [lo] or 1 [hi]) which specifies what logical state is read when the button is pressed.
- **Delay** is a variable/constant (0-255) which specifies down-time before auto-repeat in BUTTON cycles.
- **Rate** is a variable/constant (0-255) which specifies the auto-repeat rate in BUTTON cycles.

- **Bytevariable** is the workspace. It must be cleared to 0 before being used by BUTTON for the first time.
- **Targetstate** is a variable/constant (0 or 1) which specifies what state (0=not pressed, 1=pressed) the button should be in for a branch to occur.
- **Address** is a label which specifies where to go if the button is in the target state.

What is important to understand is that delay and rate refer to cycles *through* the Button command—they are not internal cycles.

The following program will make this very clear. Connect a 10 kΩ pullup between +5 V and Pin0. Connect an N.O. pushbutton switch between Pin0 and ground. Connect an LED through a 470 Ω resistor between Pin7 (anode) and ground.

```
' BUTTON Demonstration Program
'
        b0 = 0                   ' clear the workspace
Loop:   BUTTON 0,0,50,10,b0,0,NoKey
        pulsout 7, 1000          ' blip the LED if button pressed
NoKey:  pause 10                 ' wait between button checks
        debug b0                 ' show button delay/rate
        goto Loop                ' loop back through
' end of program
```

Enter and run the program.
For review:

```
BUTTON 0,0,50,10,b0,0,NoKey
       |  |  |  |  |  |  |
       |  |  |  |  |  |  |
       |  |  |  |  |  | |------ branch address
       |  |  |  |  | |-------- branch if button NOT pressed
       |  |  |  | |---------- workspace variable
       |  |  | |------------ auto-repeat rate (~100 ms) *
       |  | |-------------- auto-repeat delay (~500 ms) *
       | |------------------ pin goes low when button pressed
       |-------------------- use pin0
```

* The auto-repeat delay and rate time are calculated from delay and rate cycles multiplied by the delay time in the branch routine. They are only approximate as other commands are executed in the loop. (NOTE: the debug option will seriously impact loop timing.)

When the program is started, the debug window will show b0 having a value of zero (0). Now press and release the button. The LED will flash and you'll notice that b0 is loaded with 50, then immediately returns to 0. What you saw was the auto-repeat delay value being loaded. Since we released the button quickly, the value was returned to 0 (which means the button is not being pressed).

Now press and hold the button for a few seconds. Again you'll see the LED flash and b0 will be loaded with 50. As you hold the button down you will see that b0's value counts down with each cycle through the loop. What you're looking at is the delay value in action. If you release the button before b0 counts down to 1, you will see that it returns to 0 and nothing else happens.

Now press and hold the button. The LED will flash and b0 will start to count down from 50. Keep holding the button. You'll see that when b0 reaches a value of 1, the LED will flash again and now b0 is loaded with 10. What we've seen is the completion of the auto-repeat delay and now the auto-repeat rate is being used. So long as you hold the button down, b0 will count down from 10 to 1, and the LED will flash when b0 reaches 1. Notice that while the button is being held and b0 is not 1, the button command causes a branch to occur. When b0 reaches 1 we drop through and flash the LED. Once the button is released, we must start all over, going through the delay before getting to the rate value.

Now I admit that flashing an LED is of little value with the button command. But consider a thermostat project. Instead of flashing the LED you could change a temperature setpoint. Pressing and releasing the button would allow you to change the setpoint slowly. Large changes could be made rapidly by pressing and holding the button. None of this is unique—most household digital clocks are set in this manner. (See THRMSTAT.BAS included in BUTTON.ZIP.)

5.2.2 Generating Tones with SOUND

The parameters *note* and *duration* in the instruction SOUND specify frequency and length of the acoustic output. Tim Christopherson has measured the resulting frequencies for a ceramic resonator of 4 MHz. Table 5.2 lists the measured values.

5.2.3 Digital-to-Analog Conversion with PWM

The instruction PWM causes an analog voltage on a connected RC circuit. Figure 5.1 shows the mechanism of voltage generation.

Depending on the pulse duration, an RC circuit will generate a different voltage over the capacitor. A short pulse can generate a low voltage only. The longer the time for loading the capacitor the higher the voltage over this capacitor.

Duty is a term for periodical pulse in general. It is the relation between pulse time and period. Therefore a short pulse has the same effect as a low duty in a periodical pulse. Duty has here a value in relation to 255.

Table 5.2
Frequencies for the Instruction SOUND

				BASIC STAMP FREQUENCY TABLE FOR SOUND COMMAND					

1	93.745	29	120.250	57	167.550	85	276.30	113	787.1
2	94.490	30	121.450	58	169.950	86	282.85	114	842.8
3	95.245	31	122.700	59	172.375	87	289.70	115	906.9
4	96.015	32	124.050	60	174.900	88	296.95	116	981.6
5	96.790	33	125.250	61	177.500	89	304.50	117	1070.5
6	97.585	34	126.600	62	180.200	90	312.50	118	1175.0
7	98.390	35	127.950	63	183.000	91	320.90	119	1304.0
8	99.210	36	129.350	64	185.800	92	329.80	120	1464.0
9	100.500	37	130.750	65	188.800	93	339.20	121	1669.0
10	100.875	38	132.250	66	191.800	94	349.10	122	1939.0
11	101.750	39	133.700	67	194.900	95	359.65	123	2318.0
12	102.625	40	135.275	68	198.200	96	370.85	124	2877.0
13	103.525	41	136.750	69	201.550	97	382.75	125	3792.0
14	104.440	42	138.350	70	205.000	98	395.41	126	5560.0
15	105.450	43	140.000	71	208.600	99	409.00	127	10400
16	106.275	44	141.650	72	212.300	100	423.50		
17	107.250	45	143.350	73	216.150	101	439.15	128	
18	108.225	46	145.100	74	220.150	102	455.95		
19	109.225	47	146.900	75	224.300	103	474.10	WHITE NOISE	
20	110.225	48	148.750	76	228.600	104	493.70		
21	111.250	49	150.600	77	233.050	105	515.05	255	
22	112.300	50	152.550	78	237.700	106	538.30		
23	113.350	51	154.500	79	242.550	107	564.05		
24	114.450	52	156.550	80	247.600	108	591.75		
25	115.550	53	158.600	81	252.850	109	622.65		
26	116.695	54	160.750	82	257.950	110	657.00		
27	117.850	55	162.950	83	264.050	111	695.30		
28	119.050	56	165.200	84	270.050	112	738.40		

To get a DC voltage with only a small ripple some periods of pulse-width modulated signals are required. The parameter *cycles* defines the number of periods.

The value of the output voltage consists of 256 discrete values in a range from 0 V up to about 5 V DC. The accuracy of the output voltage cannot be better then 0.4% of the target value.

Finally we must have a look at timing. An analysis of the output voltage at the output pin with an oscilloscope during execution of the PWM instruction gives an idea of the timing. For this test a simple program will be used. The oscilloscope is connected to Pin7:

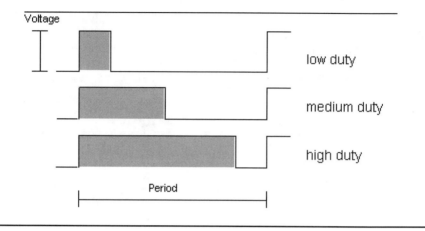

Voltage

low duty

medium duty

high duty

Period

Figure 5.1
Terms for pulse-width-modulation

```
symbol duty = ...              'Value set up before start

loop: pwm 7, duty, 255
      goto loop
```

Three different values will be used for duty and the resulting display should be stored. Figure 5.2 shows the results.

The shortest time shown in Figure 5.2 is about 50 µs and the duty results from this period. Normally the period is constant. The period for BASIC Stamps varies, too. For a duty of 3, a period of about 4 ms is required.

The execution time for this instruction can be measured in a loop. This measurement can be done with the next test program and a simple stop watch.

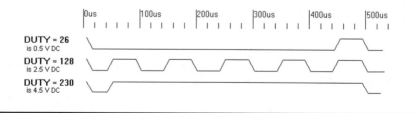

Figure 5.2
Timing for PWM Instruction

```
symbol n = w5
symbol duty  = ...        'Value set up before download
symbol cycle = ...        'Value set up before download
symbol end   = ...        'Value set up before download


sound 0,100,500           'Acoustic signal for start the watch
for n = 1 to end
      pwm 7, duty, cycle
next n
sound 0,100,500           'Acoustic signal for stop the watch
```

The parameter end defines the number of loops to be executed. This number should have a value suited to getting a comfortable measurement with a normal stop watch. The loop call needs about 1 ms for one turnaround.

From these observations, it follows that the execution time is nearly independent from the duty value itself. For all duty values there is an execution time of about 1.32 to 1.34 sec for 200 cycles. The execution time is ascertained by the number of pulse periods defined by the parameter *cycles*.

It is not possible to give a simple instruction for calculating the values for the RC circuit, so it is necessary to search for a compromise. The time constant of the RC circuit should be large enough for good suppression of the voltage ripples, and small enough to follow changes of the duty value. The compromise for most cases is a time constant of 10 ms as a product of the resistor and capacitor value. Figure 5.3 shows all of the required I/O circuitry.

Each CMOS port of the BASIC Stamp has its output resistance. Therefore the resistor value Rs = 10 kΩ is large enough to avoid a voltage division by the load. The capacitor value results from this resistor value to 1 μF.

The voltage level generated across the capacitor has a correct value only without any load. In practice there is always an input resistance of the following

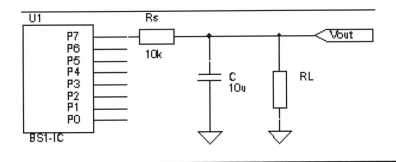

Figure 5.3
I/O circuitry for PWM

circuitry, and so the symbolic load resistance R_L will have no endless value. The real output voltage V_{out} results from voltage dividing by $R_S/(R_S + R_L)$ again.

The higher the load resistance R_L the more exact is the voltage level at V_{out}. An additional error of 0.1% minimum only comes from a load resistance of 10 MΩ in maximum.

The demands for a high input resistance of the following circuitry can be achieved by active circuits with operational amplifiers. An op-amp as non-inverting circuit or voltage follower meets the requirements in general. Figure 5.4 shows an op-amp as voltage follower with a gain of +1. It is possible to set a gain that differs from +1 with a changed feedback.

5.2.4 Resistor Measurement with POT

Both BASIC Stamps have the interesting feature that each input pin can serve as an analog input. There is only one condition: The analog value must have a resistor characteristic. In some cases this condition will be no problem.

The resistor to be measured will be calculated over a measurement of time for discharging a capacitor with a known value. Some remarks and explanations should show the limitations of this procedure and give some practical hints.

The discharge of a capacitor happens according to an exponential function. Therefore the discharge is larger in the beginning than near the end.

With a trigger circuit, it is possible to detect the specific voltage level achieved during the discharge. The resistor value can be calculated if the capacitor value, the time for discharge to the trigger level, and the voltage across the capacitor before discharge are well known. The capacitor value is defined to 100 nF and should have low leakage and a low temperature coefficient.

From the described procedure, you can see the limitations to a small resistor range. It is necessary to search for a compromise between digital resolution at the beginning of discharge and a well defined time value in the flat part of

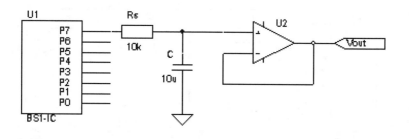

Figure 5.4
Circuit enhancement by an op-amp

the function. With the allowed resistance range from 5 kΩ to 50 kΩ only one decade of the resistance value is recordable. Many sensors, thermistors, photo resistors, etc., work in this range.

An assignment of the digital resolution to the highest resistance value is optimal for the required numeric calculations. To download and execute a scaling subroutine, key **Alt-P** from the editor. Set up the maximum resistance and look for the minimum *scale* value.

The scale values were determined for the following resistance values.

Resistance	6 kΩ	12 kΩ	30 kΩ	48 kΩ
Scale	205	115	61	52

Figure 5.5 shows the results in relation to the resistance value with fixed scale values. Figure 5.5 shows that in a range of small resistance values, preference is given to high scale values. Further, there is an enhanced sensitivity against disturbances in the upper range of resistance. Appropriate measures like short connections, twisted or shielded lines can be recommended.

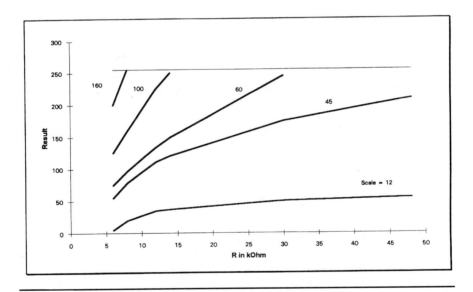

Figure 5.5
Calculated result in relation to the resistance value

5.3 Numeric Operations

PBASIC has many numerical and logical operators. Tables 5.3 and 5.4 show all operators for BASIC Stamp I and BASIC Stamp II. The size of the operands will not be checked. From PBASIC there are no format tests in a numerical and/or logical operation.

The addition problem below is an example of what happens when the valid number range is exceeded.

```
x var word
y var byte
z var word

x = $FFF0 : y = $20
z = x+y                        '  z = $10010

debug ihex4 z                  '  z = $10
```

It is easy to calculate the correct result to $10010. But only $10 was the result displayed via debug.

In the case of multiplication of two words, the result is 32 bits. PBASIC therefore has two multiplication operators for BASIC Stamp I and three for BASIC Stamp II. Three small program samples show the use of these multiplication operators. The result is 16 bits in each case, included in the program samples as a remark in the last line. The operation for all three examples is $A000 * 2 = $ 00 01 40 00. The spaces in the result separate the bytes for better orientation.

Table 5.3
Unary operators in PBASIC

Operation	Explanation	Valid for
SQR	Square-root of unsigned 16-bit value	BS2 only
ABS	Absolute of signed 16-bit value	BS2 only
~	One's complement of 16-bit value (NOT)	BS2 only
−	Two's complement of 16-bit value (NEG)	
DCD	2n decoder (0,1,2,..,15−1,2,4,..,32768)	BS2 only
NCD	Priority encoder of 16-bit value	BS2 only
SIN	Sine of 8-bit value; result is between −127 and +127	BS2 only
COS	Cosine of 8-bit value; result is between −127 and +127	BS2 only

Table 5.4

Numerical and logical binary operators in PBASIC

Operation	Explanation	Valid for
+	Add	
−	Subtract	
*	Multiply and return low 16-bits of result	
**	Multiply and return high 16-bits of result	
*/	Multiply and return middle 16-bits of result	BS2 only
/	Divide and return quotient	
//	Divide and return remainder	
MIN	Limit value to minimum	
MAX	Limit value to maximum	
&	logical AND	
|	logical OR	
^	logical XOR	
&/	logical AND NOT	
|/	logical OR NOT	
^/	logical XOR NOT	
DIG	Returns decimal digit; "12345 dig 3" returns 2	BS2 only
<<	Shift left	BS2 only
>>	Shift right	BS2 only
REV	Reverse order of bits	BS2 only

```
x con $A000        x con $A000        x con $A000
y con $2           y con $2           y con $2
z var word         z var word         z var word


z = x**y           z = x*/y           z = x*y
debug ihex4 z,cr   debug ihex4 z,cr   debug ihex4 z,cr
' $0001            ' $0140            ' $4000
```

For division, we have two operators for both BASIC Stamps. The first operation returns the quotient and the second operation the remainder. The result displayed via debug is included as a remark in the last line again.

```
x con $20               x con $20
y con $3                y con $3


z = x/y                 z = x//y
debug ihex4 z           debug ihex4 z
' $000A                 ' $0002
```

Logical operations are used in mask operations. The result of the logical operators AND, OR, and XOR are examples for the other logical operations.

```
x con %00001111     x con %00001111     x con %00001111
y con %10101010     y con %10101010     y con %10101010
z var byte          z var byte          z var byte

z = x & y           z = x | y           z = x ^ y

debug ibin8 z       debug ibin8 z       debug ibin8 z
' %00001010         ' %10101111         ' %10100101
```

The operators min and max limit a value to minimum and maximum. The kind of notation is important. There is no unique handling of these functions. The following example shows the result of this operation via debug again.

```
x var byte
y var byte
z con 5

for x=0 to 9
    y = x min z
    'y = x max z
    debug dec x," ", dec y,cr
next
```

The program text contains both functions. The max function is commented out and has no effect. The function of interest remains uncommented. The value of variable x steps from 0 to 9. Because the min function sets the variable y to 5 for all values of x under 5, the value of variable y will go from 5 to 9 only.

For BASIC Stamp II there are some further interesting operations. The function rev sorts the bits of a byte in reverse order. A program sample helps to understand.

```
x con %01100100
y var byte
i var nib

for i = 1 to 8
    y = x REV i
    debug dec i," ",ibin8 x," ", ibin8 y,cr
next
```

The output from debug is as follows:

```
1 %01100100 %00000000
2 %01100100 %00000000
3 %01100100 %00000001
4 %01100100 %00000010
5 %01100100 %00000100
6 %01100100 %00001001
7 %01100100 %00010011
8 %01100100 %00100110
```

The last line shows the resulting byte in bits, in reverse order. In the seventh line the seven LSBs will be cut and in reverse order in the result. If you cut in the third line three LSBs (100), then the result after reversing is (00000)001.

Besides the binary operators, we have for BASIC Stamp II some new unary operators. For these operators there is one common program sample.

```
x con $1000
y con %00011100
z var word
t var word

z = sqr x
debug "SQR ",dec x," = ", dec z,cr
z = ~y
debug "NOT ",ibin8 y," = ", ibin8 z,cr
z = -y
debug "NEG ",ibin8 y," = ", ibin8 z,cr
z = NCD x
debug "NCD ",dec x," = ", dec z,cr
t = DCD z
debug "DCD ",dec z," = ", dec t,cr
t = sin y
debug "SIN ",dec y," = ", dec t,cr
t = cos y
debug "COS ",dec y," = ", dec t,cr
```

The results will be displayed with the debug instruction again.

```
SQR 4096 = 64
NOT %00011100 = %11100011
NEG %00011100 = %11100100
```

```
NCD 4096 = 13
DCD 13 = 8192
SIN 28 = 81
COS 28 = 98
```

At first we get an integer square-root from a word variable. The complements of 1 and 2 are often needed operations. Both operators work with word variables, too. For calculations like bitmasking, etc., the relationship will be required often. For BASIC Stamp I it was necessary to implement a lookup table, while for BASIC Stamp II the operators DCD and NCD can help. *Attention:* The result for NCD is faulty. You have to correct the result by subtracting one.

The two last operators are the sine and cosine functions. The whole circle is described by one byte that means 256 steps. The resulting formula for a scaled sine function is shown in the second line.

$$y = \sin(x) \qquad \text{for } x = 0 \text{ to } 2\pi$$

$$Y = 127 \cdot \sin\left(\frac{X}{256} \cdot 2\pi\right) \qquad \text{for } X = 0 \text{ to } 255$$

For a better orientation both functions are shown for one period in a simple graph, Figure 5.6 below. Take an x-value, and you know immediately which value must be the result. Compare the values for sin(28) and cos(28) with the printout from debug shown above.

Finally, let us look at the calculation of a formula in PBASIC. Math has defined rules for working out, but PBASIC ignores these rules and works only from the left to the right. You have to pay attention to this feature of PBASIC while writing programs.

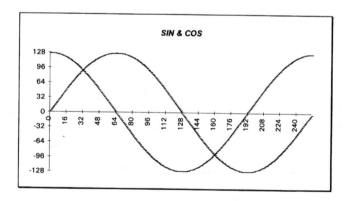

Figure 5.6
Excell chart

If you have not forgotten the arithmetic rules, you will find that the following result is correct:

$$2 + 3 * 4 = 14$$

However, calculating in this order, PBASIC's result is 20, and we have to accept this as a correct result in PBASIC's world. In PBASIC, the programmer has to write the formula as in the second or third column below:

```
'Faulty              'Correct              'For BS2 only
x = 2 + 3 * 4        x = 3 * 4 + 2         x = 2 + (3 * 4)
debug dec? x         debug dec? x          debug dec? x
' x = 20             ' x = 14              ' x = 14
```

BASIC Stamp II now permits the placing of parentheses, as the third example shows. Unary operators have the highest priority and binary operators have the second priority, and with those rules expressions are resolved left-to-right. Parentheses can override priority.

5.4 Source Text Formatting

Now you know all the instructions and directives of PBASIC for BASIC Stamp I and BASIC Stamp II, and programming first samples will be the next step.

Formatting the source texts helps for readability of programs, which is a very important point. Every programmer has asked himself—What does this mean, this work I did two weeks ago? This situation can be avoided if a program is well structured and formatted, and has enough comments.

All program samples described in this book have a uniform source text format. This program form should be considered as a proposal, and appears as follows:

```
' -----[ Title ]----------------------------------
'
' File......   .BAS or .BS2
' Purpose...
' Author....
' Started...
' Updated...

' -----[ Program Description ]------------------
'
'
```

```
'  -----[ Revision History ]---------------------------
'
'  MM/DD/YY: Version 1.0

'  -----[ Constants ]-----------------------------------
'


'  -----[ Variables ]-----------------------------------
'


'  -----[ Initialization ]------------------------------
'


'  -----[ Main Code ]-----------------------------------
'
start:

'  -----[ Subroutines ]---------------------------------
'
```

The simplest way to use this form is to always save it as FORM.BAS or FORM.BS2. You can create a new file for editing by opening this file FORM.* with an empty form instead of creating a new one. The first time you save the file, you must give it a new name. Please do not forget this renaming, because otherwise the file FORM.*, with its empty form, will be overwritten.

On the Way to Application: 6
The Practical Use of
BASIC Stamp

Each user of BASIC Stamp I or BASIC Stamp II starts with some material, his specific knowledge of electronics and programming, and more or less experience with microcontrollers, such as a BASIC Stamp. His goal is to learn by means of examples or to make applications quick and easy with them. A look at Parallax's E-Mail List Archive in the World Wide Web (URL http://www.parallaxinc.com/lists/archive-search.html) shows the variety of users and the different questions regarding practical use. So we will give some fundamental ideas and hints for using the BASIC Stamps.

6.1 Experiments and Some Tools

There are two different ways to use the BASIC Stamps:

- One is to realize a circuit or a part of a device, based on a finished development. In magazines and books, complete solutions are often described. So you can make a solid construction with all components without experiments, after you have gotten started.
- The second way is to perform certain tasks on the way to a successful project. Here you need a temporary construction like a breadboard, easy to modify and to expand with several components for the specific task. You have to work for clearness and security, for the components and for yourself. For this job we took the "main component" with connectors on a wooden board, shown in Figure 6.1.

We used the BASIC Stamp without any soldered connections on it or on the carrier boards. Plugged connectors allow for quick changes, without stress caused by soldering heat to the BASIC Stamp. All positions of connectors and important wires should be marked with names that make sense.

Figure 6.1
Experimental breadboard

The voltage from the external power supply is 12 V DC, stabilized and current-limited. To avoid a false connection between the board and the power supply, we inserted a power diode on the board, with its cathode directed to plus (+). BASIC Stamp I has an inner stabilization from the LM2636 and BASIC Stamp II from the SB-1350 HC. Both BASIC Stamps work within a wide range of supply voltage.

But we think that many attached components to BASIC Stamp work better with a stabilized supply voltage. You can use the inner stabilization of the BASIC Stamp to have a stabilized voltage of 5 V, but you must allow for a limitation of current. So we used a separate stabilization to 5 V on our breadboard with an LM 7805. Some integrated circuits need their own capacity, in the form of a ceramic capacitor of approximately 47 nF.

There are not only CMOS devices with low power consumption. There are also drivers for power components like motors and relays. A big power supply is a good accessory for the whole project. But if you start with the BASIC Stamp, your first program could be a blinking LED. In this case, a power supply with a simple battery is sufficient.

All wires should be as short as possible. This is critical in the connections of some circuits. Some time-keeper devices need a crystal with a capacity smaller than 6 pF, for example. But often it does not work, because the sum capacity of long connections prevents success.

A good help is a digital measuring device for current, voltage, resistance, and capacity. If you are lucky, you may have an oscilloscope of your own, even an older model. Many events occur in milliseconds. The possibility of trigger is important, to synchronize with pulse transients. Often we use a stop clock when we have multiple go statements or parts of programs in a loop, to get approximate time information on run-time single events.

Be aware of static discharge if you experiment. Connect yourself to a ground before handling. Don't let unused input pins go without a defined level. Pull-up or pull-down these pins with resistors. Some data lines between a BASIC Stamp and peripheral components work only with pull-down or pull-up resistors with a value of about 47 kΩ. If there are different results running the same program at different times, most undefined levels on input pins are the cause of failure. Pull-up or pull-down resistors prevent these problems.

At last we have completed our breadboard with two fixed tools: There are two LEDs with driver transistors and a key. The whole onboard circuit for BASIC Stamp II is shown in Figure 6.2.

You may complete your board with often-needed components: LED displays, piezo beeper, potentiometer, and lamps. Also, you should insert connectors like DB9, DB25, or multiple in-line screw elements.

If you do not wish to make your own experimental board, you can buy it with equipment from Parallax, too. The offered BASIC Stamp Experiment Board is designed and manufactured by HUMANsoft in Hungary. The board has a complete BASIC Stamp I built-in and has sockets for BS1-IC and BS2-IC. Figure 6.3 shows the BASIC Stamp Experiment Board from HUMANsoft.

Here we mention some additional components for BASIC Stamps offered by Parallax:

• Serial LCD Module. It is a combination of two products from Scott Edwards Electronics: Their LCD Serial Backpack and a 1x16 LC display. Connected together, they form a serially-driven LCD.

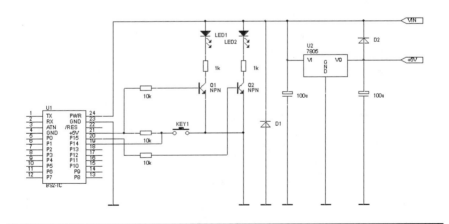

Figure 6.2
Fixed components on user board

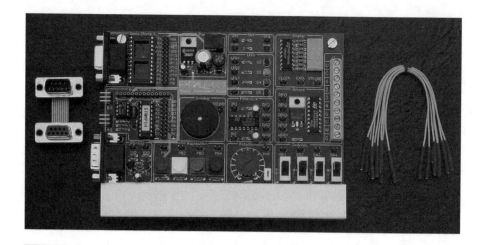

Figure 6.3
BASIC Stamp experiment board

- Stamp Stretcher. With these devices from Scott Edwards Electronics, you expand the I/O capabilities of the BASIC Stamp. This simple module adds 16 I/O lines and one analog input or 8-kByte EEPROM.

The stretcher only takes one I/O line and ground and is controlled with SEROUT instructions from the BASIC Stamp. The onboard A/D converter works in voltage measuring range 0 to 4.7 V DC with 8-bit resolution.

In this way you can expand the I/O possibilities of a BASIC Stamp I to much more than provided by BASIC Stamp II. But keep in mind, the serial transmission works more slowly.

6.2 Hints for Programming and Debugging

Our breadboard in Figure 6.2 shows two LEDs with its driver transistors. LED1 is ON after the power supply begins and the BASIC Stamp's voltage regulator is working for output V_{CC}. So you have a signal for both. LED2 is connected with the highest pin of port (Pin15). If you do not need this pin for another task, it is programmable as output for several tests.

Here is a very simple example for BASIC Stamp II:

- Blinking program
  A "test of health" of BASIC Stamp and a successful download from the PC with serial connection. It is a quickly made program:

```
loop: low 15
      pause 500
      high 15
      pause 500
      goto loop
```

- Light pulse subroutine
  Serves as a "marker" to invert temporarily in a program. So you have a tool to follow the run-time events, first of all inserted loops.

```
lamp: high 15
      pause 50
      low 15
      return
```

Modifications for BASIC Stamp I concern only the used pin numbers.

- Breakpoint
  Another pin we have used for test purposes: Pin14 is used for a given key signal. So you have the possibility for a "manual interrupt." With this key, here is the subroutine to follow:

```
brk : if in14 = 1 then brk
      debug cr,"follow the program", cr
      pause 500
      return
```

Embedded as subroutine and callable with gosub brk for a test, it is used as a breakpoint to interrupt the program while running. You can also use this key for several other purposes, for example to manually stop closed loops.

Some methods to make programs without syntactical or functional failures are given by instructions and by directives of the BASIC Stamps themselves.

1. debug
   Used throughout the program to display transient values and variables in different formats. Because debug is executed in a short time, it is efficient to couple this instruction with one of the following instructions:

```
end
pause <time>
```

2. goto <mark>
   Often we need this to jump over a failure or a part of the program that is not currently being used.

3. `gosub <mark>`

This instruction provides a temporary jump to a subroutine, for example to detect failures or to give acoustical or visual signals to mark events.

4. `toggle <pin>`

You choose the polarity of input pins to verify the conditions for working programs, so you will be sure that the program is tested for all variations of later use.

5. `end`

It is used like a breakpoint: In each part of the program, you can interrupt at run-time for tests, often with the insertion of other temporary test instructions.

6. `comment " ' "`

Besides communicating instructions we declare temporarily unused instructions with preset " ' " as comments. We jump over these instructions without deleting them.

After discussing short selfmade programs, we recommended a unique program form in Chapter 5, the section called Source Text Formatting. In the subroutine part, the test programs mentioned above should be integrated. The following listing shows this changed part.

```
'  -----[ Subroutines ]---------------------------- -------
'
lamp: high 15                    'LED driver on pin 15
      pause 1000                 '... only for test
      low 15
      pause 1000
      return


brk : if in14 = 1 then brk       'breakpoint, go on
                                   with key on pin14
                                  '...only for test
      debug cr, "follow the program",cr
      pause 500
      return
```

So you have a unique display of a program with more possibilities for find and exchange. You should make your programs as structured as possible. In this way you simplify debugging and general use.

To make subroutines is one way, but you will notice the longer run-time in critical situations. Also, keep in mind the number of subroutines. For BASIC

Stamp I there is a maximum of sixteen possible subroutines. And you can insert one subroutine in another only four times.

Also, note another restriction of BASIC Stamp I: The direct programming registers must be used with overlook reserves. Do not use register W6 if subroutines are inserted. This register works as a stack for used subroutines.

6.3 Run-Time of PBASIC Instructions

Those working with BASIC Stamps frequently ask about the possible speed of executing instructions or blocks of instructions. If you code programs in an assembler, each instruction's run-time is the sum of required machine cycles and is known by table. Using PICs the assembler instructions require a relatively short time. But instructions for BASIC Stamp are the result of interpretation of tokens from EEPROM. For users, there are two practical ways to get an approximate time for the execution of instructions or blocks of them.

6.3.1 Measurement of a Single Instruction with a Scope

For BASIC Stamp II we use a scope connected to pin15 to measure run-time for calculating the Pathagarian theorem with the sqr function. Remember that in older models of scopes, the dials in front do not show the real values. First, calibrate the time base with the crystal calibrated pulsout instruction of the BASIC Stamp. Then make the program:

```
        a var word
        b var word
        c var word

start:  a = 3: b = 4
        high 15
        c = sqr((a*a) + (b*b))
        low  15
        debug "qroot = ", dec c, cr
        goto start
```

Measuring the pulse width on the scope, we get the following values:

a	3	150
b	4	180
time	0.8 ms	1.2 ms

Measuring the time loop above without the `sqr` function gives a time of approximately 0.15 ms. We will ignore this value for calculation.

6.3.2 Measurement of Multiple Instructions with a Stopwatch

This method is based on a clamping loop with acoustical signalization of BASIC Stamp counts. The index of loop counts is fixed for each measurement. The count is varied by tests to get practical values for measuring the time difference between the two sounds. For BASIC Stamp I we use the following simple test:

```
symbol test = w5
symbol maxz = 50000

sound 0,(100,1000)    'speaker at pin 0
for test = 1 to maxz
                      'instruction or block of instructions.
next maxz
sound 0,(100,1000)
end
```

In this way we get the result of an approximate run-time for the following operations:

Operation		Execution Time
Addition	W4 = B2 + B3	0.6 ms
Multiply	W4 = B2 * B3	0.8 ms
EEPROM read	read 100, B2	0.86 ms
EEPROM write	write 5, 3	4.33 ms
Debug	debug test	220 ms
Serial input	serin 0, 2400, byte	30 ms

This chapter on the practical use of stamps will be closed with a quick mention of two remarkable tools.

The first is the matter of getting help with the function key **F1**—unfortunately only for BASIC Stamp II. This is a quick tool, used almost daily in programming.

The second tool is a little more difficult to handle, but you can get the actual knowledge and experience of Parallax, and many friends and specialists over the whole world. You can participate in this pool by use of Parallax's BBS (916-624-7101) and/or the Parallax pages in the World Wide Web (URL http://www.parallaxinc.com/stamps).

BASIC Stamp Applications 7

It is not easy to systematically describe applications for the BASIC Stamps. What principle should we use to determine how to order the information?

Remember: Both BASIC Stamps are microcontrollers with about the same high-level instruction set, but different hardware resources. The PBASIC instruction set is well-defined to solve problems in automation. The development system gives short programming cycles (edit, compile, test, debug), and the whole BASIC Stamp system meets the requirements for rapid prototyping in this class of applications. This situation puts the application in the foreground of the description.

In some cases, it will be possible to do a particular job with either the BASIC Stamp I or the BASIC Stamp II. The costs could give preference to BASIC Stamp I. In other applications, the enhanced hardware resources or new PBASIC instructions will give preference to BASIC Stamp II. All applications described here in regard to BASIC Stamp I are also suitable for BASIC Stamp II. The user can choose the Stamp that best fits the particular circumstances.

In this chapter, the application is in the foreground. We will describe why each BASIC Stamp is used. Next to the name of each application, we indicate which BASIC Stamp is used with [BS1] or [BS2].

7.1 Basics for Applications

This chapter describes some basics used in many applications. One superior feature of both BASIC Stamps is their usefulness in serial communications. Serial communications means here asynchronous serial communication according to RS232. The behavior of the pins for analog and/or digital I/O is a further point of interest.

7.1.1 [BS1]—RS232 Interface Between BASIC Stamp and PC

This chapter describes the serial communication between any I/O pins of the BASIC Stamp and the serial interface of the PC. For the connection of peripheral devices, the PC has a minimum of two serial interfaces. These interfaces are known as COM1 and COM2. Enhancements to four, eight, and more serial interfaces on the PC side are possible, but not important to this discussion. Via this interface the PC can communicate with the application program running on a BASIC Stamp. Do not confuse this interface with the serial interface for downloading.

First we need to look to the electrical signals of the serial interface of the PC. This interface does not work simply between +5 V and GND (0 V) as both BASIC stamps do, but between +12 V and −12 V, symmetrically. A logical high level is defined as +12 V, and therefore a logical low level is −12 V. The PC works internally with +5 V, too, so each PC has level converters to give the required levels to its interfaces. A well-known member of this level converter family is the MAX232 from MAXIM; it has been copied by other manufacturers, and the copies are known by the same name.

In Figure 7.1, the left side shows the logical levels of the BASIC Stamp and the right side shows the levels of the PC.

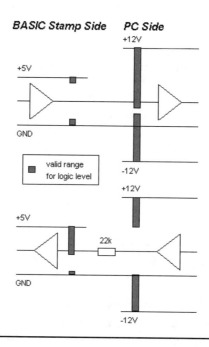

Figure 7.1
Logic levels for PC connection

The output level of a BASIC Stamp output is nearly symmetrical and is nearly equal to the rail voltage. The input side of the converter IC in the PC is not symmetrical. A high level ranges from about +1.7 V to +12 V, and a low level ranges from about −12 V to +1.2 V. The connection shown at the top of Figure 7.1 allows the PC to easily recognize the correct logical output level from BASIC Stamp.

In the other direction, the situation is not so simple. The output level of the converter IC inside the PC depends on the connected load. In Figure 7.1 this load is more then 22 kΩ. That means that the output voltage of the converter IC is out of the maximum rating of input supply voltage for BASIC Stamp. There is a simple reason that the displayed connection between the PC and the BASIC Stamp is also valid. Each I/O pin has a robust protection circuitry against electrostatic discharge (ESD) and/or electrical overstress (EOS).

Figure 7.2 shows the parts of the simplified protection circuit. From each I/O pin, diodes with a maximum clamp current of 20 mA are connected to the supply rails (V_{CC}, GND). An input voltage above V_{CC} or under GND can drive a current through the particular clamp diode to the rail. If the current flow is limited to a value smaller than 20 mA, the active clamp diode protects the I/O against a voltage level outside the valid DC characteristics. The recommended resistor value of 22 kΩ is a good choice for multiple RS232 links, too.

In Figure 7.3, the left side shows the resulting schema for a serial link between the PC and the BASIC Stamp, as described. On the right side, the "correct" interface circuit is shown. Each 5 V equipment has its own converter circuit, and between the converter circuits the communication happens with the specified logical levels.

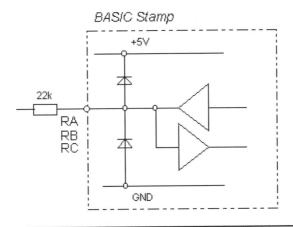

Figure 7.2
Protection circuit for an I/O pin

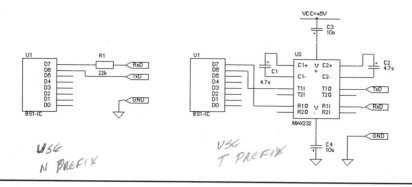

Figure 7.3
Interface circuits for RS232 links between BASIC Stamp and PC

The simple interface would be preferred for a quick design, with only a short length for the communication link.

One problem is not obvious from the figures. The converter circuits all invert the polarity of the signal. If you work with two converter circuits, the polarity will be inverted two times, and the receiver will get the correct character(s) from the transmitter. If you use the simple interface, the receiver will get all bits with inverted polarity.

The SERIN and SEROUT instructions therefore have a polarity switch in the parameter baud mode. The prefix T in the baud rate parameter signs true polarity (not inverted), and the prefix N the negated polarity (inverted).

For the left circuit in Figure 7.3, one must use the prefix N. There is only one inversion by the PC's inverter; that's why the software on BASIC Stamp's side has to take a negation, too. For the right circuit in Figure 7.3, the prefix T is the right choice because two inversions are performed by the two converter circuits. After the description of the hardware conditions for the serial communication between a BASIC Stamp and the PC, we have to clarify the software requirements on the PC's side.

Communication with a Monitor Program   In Chapter 4, the section called Programs for Serial Communication, two programs were demonstrated that can each serve as a simple monitor program. Such a program must send any characters to the connected BASIC Stamp, and must receive all characters sent by the BASIC Stamp.

It is important for the communication parameters for the COM interface of the PC and the BASIC Stamp to be same.

For a first test the simple interface without the MAX232 is good enough. The TxD line of the chosen COM port will be connected with the RxD line of the BASIC Stamp (Pin for SERIN). In the same manner, the RxD line of the

COM port will be connected to the TxD line of the BASIC Stamp (Pin for SEROUT). GND is connected directly. This simple three-wire interface is the base of many communication links.

The handshake lines of the chosen COM port are not connected if the PC software ignores their potential. If the software asks for the potential of the handshake lines, you can outwit the PC by shortcircuiting the marked pins, as shown in Figure 7.4.

After the electrical connections are made, the BASIC Stamp must have a test program for communication with the monitor program. The following test program for a BASIC Stamp I receives one character at Pin7 and shows it as a hex number in the debug window. Next, the code of the received character will be incremented and sent back via Pin6. In other words, if the BASIC Stamp receives the character "A," it will send a "B" back to the monitor program, and so on.

```
       symbol RxD  = 7
       symbol TxD  = 6
       symbol baud = N2400
       symbol char = b0

       debug cls
loop:  serin RxD,baud,char    ' receive one char
       debug $char
```

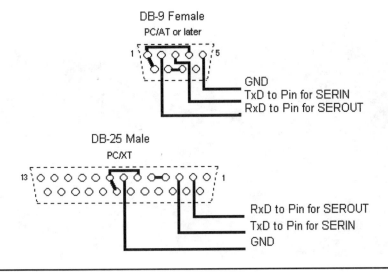

Figure 7.4
Connection of BASIC Stamp to the PC's COM port

```
char = char + 1          ' increment char
serout TxD,baud,(char) ' transmit one char
goto loop
```

The serial interface works with 2400 Baud. The prefix N signalizes the software negation because the simple interface has only one inverting converter circuit.

The parameters at both sides of the serial links are the same as those shown in the monitor program in Figure 7.5. All conditions for a correct communication should be given.

The test program for BASIC Stamp I contains the debug instruction to show the hex code of the received character. The use of the debug instruction means that the BASIC Stamp development system is running on the PC. Because it is not possible to run the monitor program at the same time on the same PC, you need a second PC to run the monitor program. Figure 7.6 demonstrates the situation.

If you want to avoid using two PCs, you must comment the debug instruction in the test program for BASIC Stamp I. Now the running test program gives no message back to the development system. You need it only for downloading. After downloading, the BASIC Stamp I runs autonom and communicates via serial link. For communication with the PC you must start the monitor program, which transmits and receives characters to and from BASIC Stamp I. Figure 7.7 shows this configuration.

The procedure described in this chapter is good for the debugging of subroutines and for first tests. In an application, the BASIC Stamp has to

Figure 7.5
Communication between RS232MON and BASIC Stamp

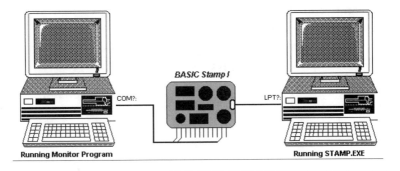

Figure 7.6
A second PC is needed for usage of the debug instruction

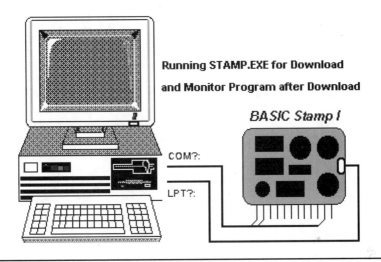

Figure 7.7
Only one PC is needed without usage of the debug instruction

talk with an application program on the PC's side, too. In a measurement system, for example, a BASIC Stamp can build the interface to a sensor. The application program running on the PC then sends commands to the sensor interface and the sensor answers with measurement values. The next two sections describe application programs for a DOS and a Windows environment.

Communication with a DOS Application   If we live with the BASIC Stamp in a BASIC world, it makes sense to program the DOS application in BASIC, too. Everybody who has installed MS-DOS on his PC has QBASIC (GW-BASIC in earlier versions) in the MS-DOS bundle.

With the same format restrictions it is possible to build a simple test program for the PC side. The following listing shows an example of such a test program.

```
' ------[ Title ]----------------------------------------------
'
' File...... RS232QB.BAS
' Purpose... To demonstrate the communication between QBASIC and
'            BASIC STAMP
' Author.... Claus Kuehnel
' Started... 25.09.94
' Updated... 25.09.94

' -----[ Program Description ]---------------------------------
'
' This program reads a keystroke and sends this character and a CR
' to the BASIC STAMP. If this is done the program is waiting for
' one character followed by a CR from BASIC STAMP. Both characters
' are displayed on the screen for inspection of a correct
' commmunication.

' -----[ Revision History ]-----------------------------------
' 25.09.94 Version 1.0

' -----[ Constants ]------------------------------------------
'

' -----[ Variables ]------------------------------------------
'
DEFSTR A-X

' -----[ Initialization ]-------------------------------------
'
Init:   OPEN "COM1:2400,N,8,1,CD0,CS0,DS0,OP0,RS" FOR RANDOM AS #1

' -----[ Main Code ]------------------------------------------
'
Main:   CLS
        PRINT "Communication via COM1"
        PRINT
        PRINT "Abort by ESC";
        LOCATE 5, 1
        PRINT "Input one character from keyboard:"
        PRINT
Again:  tra = INKEY$
        IF tra = "" THEN GOTO Again
        IF tra = CHR$(27) THEN GOTO Done
```

```
            PRINT #1, tra
            INPUT #1, rec
            PRINT "Character transmit:"; tra; "Character received: "; rec
            GOTO Again
Done:       PRINT : PRINT
            PRINT "Communication stopped."
            PRINT
            CLOSE
            SYSTEM

' ----[ Subroutines ]-------------------------------------------
'
```

The program starts with the initialization of the used serial port, COM1, with a baud rate of 2400 Baud, no parity, 8 databits, and 1 stop bit. The remaining parameters switch off the handshake signals because we work only with a simple three-wire interface. If you want to change the used COM port you have to make changes in the source text.

After outputting some messages to the screen, the program queries the keyboard. If the character input by the keyboard was not ESC, then this character and an attached CR will be transmitted via COM1 to the BASIC Stamp (PRINT#1, tra). Immediately after transmitting, the COM port waits to receive two characters from BASIC Stamp (INPUT#1, rec). Before the next keyboard query, both characters—transmitted and received—will be displayed in the screen. An input of the character ESC will stop the program after closing the communication channel.

From the QBASIC listing we can see that the instructions PRINT#1, tra and INPUT#1, rec, respectively, were responsible for transmitting and receiving one character. The instruction PRINT#1, tra transmits not only that character behind the variable tra but also a carriage return CR ($0D).

The small test program for BASIC Stamp I must be adapted to this condition as follows:

```
            symbol RxD  = 7
            symbol TxD  = 6
            symbol baud = N2400
            symbol char = b0
            symbol eom  = b1

            debug cls
loop:       serin RxD,baud,char, eom      ' receive one char and CR
            debug $char,$eom
            char = char + 1               ' increment char
            serout TxD,baud,(char,eom)    ' transmit one char and CR
            goto loop
```

For the attached CR a new variable `eom` was implemented. The CR is saved only temporarily and will be sent back immediately to fulfill the requirements of the instruction `INPUT#1,rec`.

The start of the PC program `RS232QB.BAS` occurs from QBASIC inside. At the DOS command line, you have to key in the underlined commands, and then press Enter.

```
C:\DOS>qbasic rs232qb <Enter>
```

Figure 7.8 shows the QBASIC environment with the loaded program `RS232QB.BAS`. To start the program, open a pull-up menu or press the function keys **Shift-F5.** Figure 7.9 shows this part of the procedure. The author's German program version of QBASIC should not give additional problems.

Communication with a Windows Application Windows is an event-controlled programming environment. This feature has an essential influence on the programming technique, independent of the programming language used. In our case we want to stay in a BASIC world, so Visual BASIC for Windows is our preferred programming environment.

In the applications following these basics, a Visual BASIC example is described in detail. In this chapter only three subroutines demonstrate the required procedure.

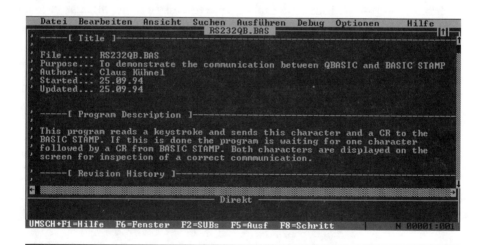

Figure 7.8
Edit window with loaded application program RS232QB.BAS

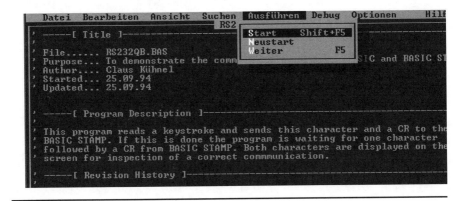

Figure 7.9
Start of application program RS232QB.BAS

Starting a Visual BASIC program means opening a first window by a subroutine Form_Load(). In this subroutine all initialization stuff can be placed.

```
Sub Form_Load ()
     ' Use COM1.
     Comm1.CommPort = 1
     ' 2400 baud, no parity, 8 data, and 1 stop bit.
     Comm1.Settings = "2400,N,8,1"
     ' Read the entire buffer when Input is used.
     Comm1.InputLen = 0
     ' Open the port.
     Comm1.PortOpen = True
End Sub
```

In our example, we work with COM1 and the following parameters: no parity, 8 data bits, and 1 stop bit. In the last step, the COM1 port was opened.

A click to the command button Command3D1 transmits the characters "RD" (Read) to a BASIC Stamp. The subroutine waits now until at least 2 characters from the connected BASIC Stamp are received. The entire buffer contents are saved in the variable InString$ for further processing.

```
Sub Command3D1_Click ()
     ' Send acommand to the BASIC Stamp.
```

```
          Comm1.Output = "RD"
          ' Wait for data to come back to the serial port.
          Do
                Dummy = DoEvents()
          Loop Until Comm1.InBufferCount >= 2
          ' Read a data word in the serial port.
          InString$ = Comm1.Input
      End Sub
```

To stop the application program, use the **Quit** button. A click to this button must close the COM1 port and end the program. The subroutine Quit_Click() carries out these two functions.

```
Sub Quit_Click ()
      ' Close the serial port.
      Comm1.PortOpen = False
      End
End Sub
```

This chapter contains no program for any BASIC Stamp because only some basics of Visual BASIC for Windows could be mentioned. Hints to the graphic design are not included at all.

7.1.2 [BS1]—Communication Between BASIC Stamps

If the BASIC Stamp can communicate serially with a PC, it should be possible for two BASIC Stamps to communicate with each other. The electrical connections are very simple. Figure 7.10 shows two BASIC Stamps connected together.

In both BASIC Stamps, Pin0 is the serial input and Pin1 the serial output.

The activities will be distributed so that the BASIC Stamp marked as U1 gets the master function and the one marked as U2 works as the slave. At Pin2 of the slave is an LED connected to signalize the activities of the slave.

The job for both BASIC Stamps is very simple; an example will illustrate. The master transmits a command byte ("1" or "0") to the slave, which evaluates the received command byte. The slave sets Pin2 to High if the received byte was a "1." If it was a "0," it resets Pin2 to Low. The connected LED displays the state of Pin2 immediately.

So the master will know whether the action is carried out, the slave gives a return value to the master. This return value is generated by reading Pin2 back. If the command from the master was not defined (whether "0" nor "1"), then the slave takes no action and the return value is set to FF_H.

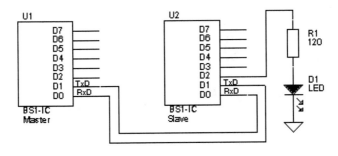

Figure 7.10
Communication between BASIC Stamps

The following two listings contain the programs for master and slave based on BASIC Stamp I.

```
' -----[ Title ]----------------------------------------------------

'
' File......  COMM_M.BAS
' Purpose...  To demonstrate the communication of two BASIC Stamps
' Author....  Claus Kühnel
' Started...  30.09.94
' Updated...


' -----[ Program Description ]--------------------------------------
'
' Two BASIC Stamps are connected over RS232.
' The master stamp sends a command to switch an output of the slave.
' After switching the slave reads this output an gives the state
' back to the master.


' -----[ Revision History ]----------------------------------------
'
' 30.09.94: Version 1.0


' -----[ Constants ]-----------------------------------------------
'
      symbol RxD = 0          ' RS232 input
      symbol TxD = 1          ' RS232 output

      symbol baud = N2400     ' 2400 Baud
```

```
' -----[ Variables ]-------------------------------------------------
'

        symbol retval  = b0        ' return byte from the slave
        symbol command = b1        ' command byte for the slave


' -----[ Initialization ]---------------------------------------------
'


' -----[ Main Code ]-------------------------------------------------
'

        pause 100
start:command = "1"               ' send command byte
        gosub send

        command = "0"             ' send another command byte
        gosub send

        goto start                ' repeat endless


' -----[ Subroutines ]-----------------------------------------------
'

send:  serout TxD, baud, (command)   ' send the command to the slave
       serin  RxD, baud, retval      ' look for return byte
       'debug retval
       pause 1000                    ' wait a little bit
       return
```

```
' -----[ Title ]-----------------------------------------------------
'

' File......  COMM_S.BAS
' Purpose...  To demonstrate the communication of two BASIC Stamps
' Author....  Claus Kühnel
' Started...  30.09.94
' Updated...


' -----[ Program Description]----------------------------------------
'

' Two BASIC Stamps are connected over RS232.
' A master stamp sends a command to switch an output of the slave.
' After switching the slave reads this output an gives the state
' back to the master.


' -----[ Revision History ]------------------------------------------
'

' 30.09.94: Version 1.0
```

```
' -----[ Constants ]---------------------------------------------

      symbol RxD = 0              ' RS232 input
      symbol TxD = 1              ' RS232 output
      symbol OUT = 2              ' pin for connecting a LED
      symbol baud = N2400         ' 2400 Baud


' -----[ Variables ]---------------------------------------------

      symbol retval = b0          ' return byte to the master
      symbol command = b1         ' command byte from the master


' -----[ Initialization ]---------------------------------------------



' -----[ Main Code ]---------------------------------------------

start:serin RxD, baud, command  ' look for a command from the master
      if command = "1" then PinHi    ' set pin Hi
      if command = "0" then PinLo    ' set pin Lo
      goto error                  ' command unvalid

PinLo:low OUT                     ' set pin Lo
      retval = pin2               ' read pin
      goto send                   ' send state back

PinHi:high OUT                    ' set pin Hi
      retval = pin2               ' read pin
      goto send                   ' send state back

error:retval = $FF                ' write errorcode to return byte

send: serout TxD, baud, (retval)

      goto start                  ' repeat endless


' -----[ Subroutines ]---------------------------------------------

```

From the listings, one can see that the BASIC Stamps communicate in a half-duplex procedure. This sounds more complicated than it is. What does half-duplex mean?

A program running on a BASIC Stamp works simply and straightforward in an endless loop. The BASIC Stamp can do only one activity at a time. For the serial interface, this means that the BASIC Stamp can transmit or receive, but cannot do both together. If two BASIC Stamps are connected via two wires for serial communication, then only one wire is active at each point in time.

In BASIC Stamp I applications, the number of I/O pins can be a limitation. In this case it could be useful to reduce the use of pins to the essential. The next paragraphs will demonstrate communication via one wire.

In the last two listings there are only small changes needed. The communication happens only via Pin0. That means:

1. Change `symbol RxD = 0` into `symbol RTxD = 0`
2. Change `symbol TxD = 1` into `' symbol TxD = 1` *Pin1 will be free*
3. Change RxD and TxD in all instructions to RTxD
4. Add one pull-up resistor from Pin0 to V_{CC} *explanation follows*

For the communication link to work reliably, some hints are required. Figure 7.11 shows a fictive timing for an inverting baud mode.

The master must switch to receiving immediately after its transmission. This preparation for receiving is required to avoid missing a character from the response of the slave.

The slave transmits its return value first after its operation so that during the period marked (1) in the timing diagram, the I/O pins of both BASIC Stamps are defined as input and the data line is in an undefined state. The problem is that each input floated, and any edge on the data line could be interpreted as a startbit of a new data transfer. The synchronization of both BASIC Stamps is lost and the whole system hangs.

A simple solution for this problem is to add a pull-up resistor between Pin0 and V_{CC} (for an inverted baud mode). In this time, when both I/O pins are defined as input, they see a well-defined line with a Hi level.

Another more complex solution is a changed timing, as Figure 7.12 shows.

The idea of the changed timing is that each BASIC Stamp sets the data line to Hi after receiving. If the transmitter changes to a receiver, it finds no floating

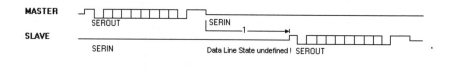

Figure 7.11
Timing for a serial communication via one connection

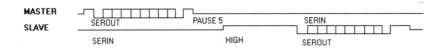

MASTER			
	SEROUT	PAUSE 5	SERIN
SLAVE	SERIN	HIGH	SEROUT

Figure 7.12
Changed timing for reliable data transfer

line but a line set to Hi. The required `pause` instructions are short, but limit the transfer speed for additional data exchange.

7.1.3 Electrical Pin Conditions for Analog and Digital I/Os

Serial communication is not the only thing possible with an I/O pin. The driving capabilities of I/O pins are also important for microcontrollers. The design of drivers for external logic and/or external loads depends on the electrical behavior of the I/O circuitry.

If the relevant pin is defined as an input, then the electrical conditions given in Table 2.5 must be fulfilled for reliable detection of the logic levels. If the pin is defined as an output, there are enhanced possibilities, depending on the concrete application.

If external logic connected to a pin is configured as an output, the electrical condition given in Table 2.5 must also be fulfilled. Basically this statement is valid if external logic and external loads together are connected to one pin. If a pin has to drive an external load with no requirements for keeping defined levels, then there is room for some variations. In such a case only the overall power conditions are important.

Consider the current-voltage characteristic of an I/O driver in Microchip's data sheet, which gives values for the output resistance of the MOS transistors in their linear region like those shown in Table 7.1. The output resistance of the p-channel MOSFET connected to the power supply V_{CC} must be higher than that of the n-channel MOSFET connected to the ground. If there are loads with a low resistance to drive, it is better to connect them between an I/O pin and V_{CC}. The internal voltage drop is thereby minimized.

Table 7.1
Output resistance on an I/O pin

Temperature	−40°C	25°C	85°C
R_{OL} at V_{DD} = 5 V	18 Ω	33 Ω	66 Ω
R_{OH} at V_{DD} = 5 V	42 Ω	94 Ω	500 Ω

The output resistance of the I/O driver should not be ignored, but its temperature dependence is even more important. A simple example shows the importance of this phenomenon. An LED is connected between an I/O pin and the ground. The required resistors for current limitations will be calculated with the next formula.

$$R_V = \frac{V_{DD} - V_{LED}}{I_{LED}} - R_{OH}$$

With an ambient temperature of 25°C, and the following values: V_{DD} = 5 V, V_{LED} = 1.5 V, and I_{LED} = 10 mA, one gets a resulting resistor of about 256 Ω.

With this resistor value and the supposition of no temperature dependence of the voltage across the LED, the changes in the LED current and in LED brightness can be calculated.

$$I_{LED} = \frac{V_{DD} - V_{LED}}{R_V + R_{OH}}$$

The LED current remains about 11.7 mA at about −40°C. With a temperature of about 80°C, this current reduces to 4.6 mA. That means, in other words, a reduction of LED brightness by about half.

Circuits built for room temperature or a reduced temperature range only must not take these problems into consideration. In industrial designs, the consideration of all influences is a must.

7.2 More Complex Applications

7.2.1 BASIC Stamp Network

A very important feature of both BASIC Stamps is their possibility for serial communication. Not only point-to-point communication, but also networking (for some BASIC Stamps) are possible for systems with distributed resources.

In the network solution described in the next two sections, we assume that one node of the network acts as a master and the other nodes act as slaves. The master can be the same BASIC Stamp as that used for the slaves or it can be a PC running an appropriate master program.

[BS1]—Master Built by BASIC Stamp   The simplest network is built with one BASIC Stamp as master and some BASIC Stamps as slaves. Figure 7.13, for example, shows a BASIC Stamp network with one master and two slaves.

On all BASIC Stamps, Pin0 is the serial input and Pin1 is the serial output. The serial inputs of the slaves will be driven from the transmit output of the master: The serial input of the master, however, will be driven from the (two) outputs.

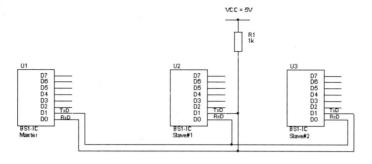

Figure 7.13
BASIC Stamp network

To avoid electrical problems, the outputs of the slaves must be Open-Drain outputs. The result is a wired-OR of all slave outputs.

For a demonstration, the following circumstances will be assumed:

1. The master sends an addressed command to a slave.
2. The addressed slave carries out the required action. The other slaves ignore that command.
3. After the action has been carried out by the slave, the slave sends a return value to the master.
4. This return value can be utilized by the master.

The next listing shows the master program for BASIC Stamp I. The two connected slaves have the addresses "A_Node" and "B_Node." An additional command byte starts a function in the addressed slave. In this example, only the command bytes "0" and "1" are allowed.

After sending a command to a slave the master waits until an answer from the slave comes back. The debug command in the debug window shows when the master has received the return value. The PC and the master must be connected via a download cable for an inspection of the return values.

The end of the loop will be shown by the character ">" on the debug screen.

```
' -----[ Title ]----------------------------------------------------
'
' File......  MASTER.BAS
' Purpose...  Demonstration of a BASIC Stamp network
' Author....  Claus Kühnel
' Started...  8.10.94
' Updated...
```

```
' -----[ Program Description ]-------------------------------------
'
' In this demonstration three BASIC Stamps build a network.
' The master sends commands to the two slaves named A_Node and
' B_Node. After execution the required function the slave sends a
' return value back to the master. This return value is displayed
' by DEBUG in this demonstration.


' -----[ Revision History ]----------------------------------------
'
'  8.10.94: Version 1.0


' -----[ Constants ]-----------------------------------------------
'
      symbol RxD = 0
      symbol TxD = 1
      symbol LED = 2
      symbol baud = T2400


' -----[ Variables ]-----------------------------------------------
'
      symbol retval = b0
      symbol command = b1


' -----[ Initialization ]------------------------------------------
'


' -----[ Main Code ]-----------------------------------------------
'
start:  command = "1"
      serout TxD, baud, ("A_Node",command) ' send "1" to A_Node
      serin RxD, baud, retval              ' wait for return value
      debug $retval
      pause 500

      serout TxD, baud, ("B_Node",command) ' send "1" to B_Node
      serin RxD, baud, retval              ' wait for return value
      debug $retval

      command = "0"
      serout TxD, baud, ("A_Node",command) ' send "0" to A_Node
        serin RxD, baud, retval            ' wait for return value
      debug $retval
```

```
serout TxD, baud, ("B_Node",command)  ' send "0" to B_Node
serin RxD, baud, retval                ' wait for return value
debug $retval

debug ">"
goto start                             ' repeat endless
```

```
' -----[ Subroutines ]-------------------------------------------
'
```

The next two source programs show what the slaves have to do. The whole job for each slave is to switch an output to Hi or Lo. The connected LED signalizes the logical state Hi with LED on, and the logical state Lo with LED off. For simplicity, the command bytes serve as return values.

Both slave programs have one peculiarity in the range Constants. Serial input and serial output work with different baud modes. On the transmitter side the baud mode OT2400 controls the driver to work as an open-drain output. A pull-up resistor pulls the output to the logical state Hi if all outputs are inactive.

```
' -----[ Title ]------------------------------------------------
'
' File......  SLAVE1.BAS
' Purpose...  Communication in BASIC Stamp Network - Slave #1
' Author....  Claus Kühnel
' Started...   8.10.94
' Updated...

' -----[ Program Description ]----------------------------------
'
' In this demonstration three BASIC Stamps build a network.
' The master sends commands to the two slaves named A_Node and
' B_Node. After execution the required function the slave sends a
' return value back to the master.

' -----[ Revision History ]------------------------------------
'
'   8.10.94: Version 1.0

' -----[ Constants ]--------------------------------------------
'
    symbol RxD = 0
    symbol TxD = 1
    symbol LED = 2
    symbol baud_in  = T2400
    symbol baud_out = OT2400
```

```
' -----[ Variables ]---------------------------------------------
'
      symbol command = b0


' -----[ Initialization ]----------------------------------------
'
      low LED


' -----[ Main Code ]---------------------------------------------
'
start:  serin RxD, baud_in, ("A_Node"), command
      if command = "1" then LED_ON
      if command = "0" then LED_OFF
      goto start

LED_ON: high LED
      serout TxD, baud_out, (command)
      goto start

LED_OFF:low LED
      serout TxD, baud_out, (command)
      goto start


' -----[ Subroutines ]-------------------------------------------
'


' -----[ Title ]-------------------------------------------------
'
' File......  SLAVE2.BAS
' Purpose...  Communication in BASIC Stamp Network - Slave #2
' Author....  Claus Kühnel
' Started...   8.10.94
' Updated...


' -----[ Program Description ]-----------------------------------
'
' In this demonstration three BASIC Stamps build a network.
' The master sends commands to the two slaves named A_Node and
' B_Node. After execution the required function the slave sends a
' return value back to the master.


' -----[ Revision History ]-------------------------------------
'
'  8.10.94: Version 1.0
```

```
' -----[ Constants ]-----------------------------------------------
'
        symbol RxD = 0
        symbol TxD = 1
        symbol LED = 2
        symbol baud_in  = T2400
        symbol baud_out = OT2400


' -----[ Variables ]-----------------------------------------------
'
        symbol command = b0

' -----[ Initialization ]-----------------------------------------
'
        low LED

' -----[ Main Code ]-----------------------------------------------
'
start:  serin RxD, baud_in, ("B_Node"), command
        if command = "1" then LED_ON
        if command = "0" then LED_OFF
        goto start

LED_ON: high LED
        serout TxD, baud_out, (command)
        goto start

LED_OFF:low LED
        serout TxD, baud_out, (command)
        goto start

' -----[ Subroutines ]---------------------------------------------
'
```

[BS1]—Master Built by PC The network described in the last section was built from identical microcontrollers for master and slave. The BASIC Stamp has its advantages in all parts of the actual process.

In this example, a PC is the network master to control a BASIC Stamp network from a Windows application program. Figure 7.14 shows a network consisting of some BASIC Stamps as slaves and a PC as the network master.

To get an attractive user interface, the master program for the PC was programmed in Visual-BASIC 3.0. In version 3.0, Visual-BASIC supports the serial interface of the PC in a simple manner. There are no additional problems due to the communication with the BASIC Stamps.

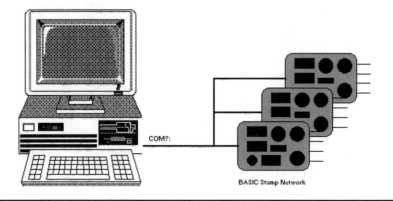

Figure 7.14
BASIC Stamp network with PC as master

The BASIC Stamp is the same in networks as it is alone. The outputs of the BASIC Stamp must be connected—wired or with a pull-up resistor to the supply voltage.

As in every communication link, the protocol for all the messages transferred is critical. In a simple point-to-point link, security is not central. In a network we always have to watch for unavoidable disturbances, and we have to search for secure mechanisms to prevent transmission failures.

In microcontroller networks, standardization of some protocols happened in the past. All procedures have a certain overhead that ties a part of the resources of the microcontroller. The BASIC Stamp has limited resources only and does not need additional overhead.

The task was to define a special protocol that takes care of the resources and the security issues.

Protocol for Data Transfer in a BASIC Stamp Network. The data transfer in our BASIC Stamp network is based on five byte messages secured by a CRC (Cyclic Redundancy Check). Details of the CRC are described in this section, because the use of a CRC to provide security in any communication and for memory tests has become widespread.

Basically, the master sends a message to all slaves. This message contains a command byte with an address and function code followed by four data bytes. All slaves receive this message and, by means of the CRC, test whether the data transmission was correct.

If a slave with the specified address is in the network, then it will interpret this message and execute the function if the result of the CRC was okay. If the CRC indicates failure, then this slave will give an error code and send the CRC back to the master. All other slaves ignore this message.

BASIC Stamp

The master waits for the reaction of the slave and evaluates that reaction. If there is no answer after a certain time, the master goes in timeout. This timeout is a hint that there is no slave with the specified address in the net. Figure 7.15 shows a structogram of a slave's reaction to a message sent from the master. The structogram thoroughly describes the handling of messages from a master to the slaves. The contents of such a message must be considered separately.

All messages contain five bytes. Table 7.2 shows the possible messages: To send a message, set the command byte and the two data bytes. The CRC bytes will be calculated and attached by the sending controller. In Bit6 to Bit4 the address of the node is coded, as shown in Table 7.3. The function to be executed by the slave is coded in Bit3 to Bit0. Bit7 serves as an error flag and can be set only by the slave.

Under these conditions, a maximum of eight slaves can build a network. Each slave can execute a maximum of sixteen functions. The program IO-CTRL.BAS is an example of a slave node based on BASIC Stamp I. The BASIC Stamp slave node with its external parts is shown in Figure 7.16.

Keep in mind that there is only one pull-up resistor at the TxD line, no matter how many slaves are connected. The LED is a status display: It brightens after receiving a valid command, until it sends the answer.

The following listing (IO-CTRL.BAS) is the source text for the slave in Figure 7.16, with a node address 7. If you want to integrate further nodes in the network, only small changes are required. In this listing the changes required are bold.

At first the new slave must get a new address, one not already in the net. Each node address may exist only once in the net, to avoid having multiple answers from identically-named slaves, resulting in communication errors.

Next the executable functions must be defined.

```
' -----[ Title ]-------------------------------------------------
'
' File......  IO-CTRL.BAS
' Purpose...  I/O Controller for Stamp Network
' Author....  Claus Kühnel
' Started...  17.06.95
' Updated...


' -----[ Program Description ]----------------------------------
'
' The I/O controller receives a three byte command plus two bytes
' CRC from a host system. After receiving the command some tests
' are taken. At first the CRC over these five bytes will be
' calculated. If the calculated CRC is equal to zero the
' transmission was without any error. The next test looks for a
' valid command (node & function). The functions of the I/O
```

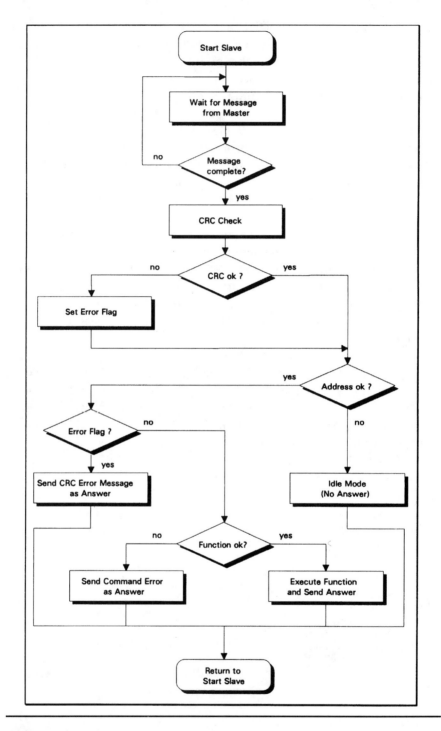

Figure 7.15
Structogram for message handling by slave

Table 7.2
Data contents in network messages

Message	COMM	Hi_Data	Lo_Data	Hi_CRC	Lo_CRC
Command from Master	00..7F	00..FF	00..FF	00..FF	00..FF
Answer from Slave (with no error)	00..7F	00..FF	00..FF	00..FF	00..FF
...at CRC Error	8X..FX	00..FF	00..FF	00..FF	00..FF
...at Command Error	C0	00..FF	00..FF	00..FF	00..FF

Table 7.3
Coding of courier and byte

B7	B6	B5	B4	B3	B2	B1	B0
0	N2	N1	N0	F3	F2	F1	F0
		Node address			Function number		

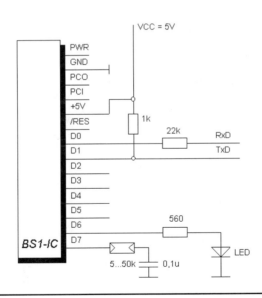

Figure 7.16
BASIC stamp node for light measurement with photo sensitive resistor

```
' controller are programmed following the labels F0, F1 and F2 for
' this example. After execution of one function the I/O controller
' transmit a five byte result back to the host.

' -----[ Revision History ]-----------------------------------------
'
' 17.06.95: Version 1.0

' -----[ Constants ]------------------------------------------------
'
      symbol RxD      = 0           ' RS232 output
      symbol TxD      = 1           ' RS232 input
      symbol LED      = 6           ' LED output
      symbol AI       = 7           ' Analog Input
      symbol baud     = N2400       ' Baudrate
      symbol CRCPOLY = $1021        ' CRC-CCITT
      symbol scale    =    93       ' Scale Factor for POT
      symbol this_node = $07        ' Node Address for this Node
      bsave                         ' (D6-D4 = 000-111)

' -----[ Variables ]------------------------------------------------
'
      symbol CRC  = w0              ' CRC Word
      symbol Lo_CRC = b0
      symbol Hi_CRC = b1
      symbol Byte = w1              ' Word Variable
      symbol Data = w2
      symbol Lo_Data = b4
      symbol Hi_Data = b5
      symbol Command = b6           ' Command Byte
      symbol Node     = b7          ' Node Nibble
      symbol Fct      = b8          ' Function Nibble
      symbol i        = b9          ' temp. Variable
      symbol Lo_Tmp   = b10
      symbol Hi_Tmp   = b11

' -----[ Initialization ]-------------------------------------------
'
      Data = 0
      High LED                      ' LED Off

' -----[ Main Code ]------------------------------------------------
'
start:' read five bytes from host
      serin RxD,baud,Command,Hi_Data,Lo_Data,Hi_Tmp,Lo_Tmp
```

```
        CRC = 0                              ' Calculation of CRC
        Byte = Command
        gosub calc_CRC
        Byte = Hi_Data
        gosub calc_CRC
        Byte = Lo_Data
        gosub calc_CRC
        Byte = Hi_Tmp
        gosub calc_CRC
        Byte = Lo_Tmp
        gosub calc_CRC


        if CRC <> 0 then CRC_error      ' Test of CRC
        goto Exec


CRC_error:
        Command = Command | $80         ' CRC Error: Set MSB in Command
        Hi_Data = Hi_CRC
        Lo_Data = Lo_CRC


Exec:Node = Command/16 & $07            ' Node Address in Hi_Nibble
        if Node <> this_node then Node_Idle ' this is Node# 07H
        if Command > $7F then Send      ' CRC-Error
        Fct = Command & $0F             ' Function in Lo_Nibble
        if Fct = 0 then F0
        if Fct = 1 then F1
        if Fct = 2 then F2
        Command = $C0                   ' Command Not Valid: Command = $C0
Send: CRC = 0                           ' Calculation of CRC


        Byte = Command
        gosub calc_crc
        serout TxD,baud,(Byte)


        Byte = Hi_Data
        gosub calc_crc
        serout TxD,baud,(Byte)


        Byte = Lo_Data
        gosub calc_crc
        serout TxD,baud,(Byte)
        serout TxD,baud,(Hi_CRC)
        serout TxD,baud,(Lo_CRC)
```

```
        high LED             ' LED Off

        goto start           ' Go back to Start and Wait for a new Command

Node_Idle:
        sleep 1
        goto start

FO:     low LED                    ' LED On
        Hi_Data = 0                ' Function# 0
        pot AI,scale,Lo_Data
        debug lo_data
        goto Send

F1:     low LED                    ' LED On
        Hi_Data = 0                ' Function# 1
        Lo_Data = Scale
        goto Send

F2:     low LED                    ' LED On
        Data = Data * 2            ' Function# 2
        goto Send

' no more functions today

' -----[ Subroutines ]-----------------------------------------
'
calc_crc:
        CRC = Byte*256 ^ CRC
        for i=0 to 7
            if Bit15 = 0 then shift_only
            CRC = CRC * 2 ^ CRCPOLY
            goto nxt
shift_only: CRC = CRC * 2
nxt:    next
        return
```

In the listing above, three functions were defined:

F0 Light measurement with photo sensitive resistor,

F1 Query calibration constant `scale`, and

F2 Data manipulation (multiplication of the received data value by two, for example)

At this time, additional functions are not defined. Depending on the available memory, further functions can be defined.

Due to the restricted resources of BASIC Stamp I, the memory may run short. At the end of the declaration part for constants, the bsave instruction saves to disk an object file for inspection with the program STMPSIZE. The particular memory dump looks like this:

```
BASIC Stamp Program Size
-----------------------
User memory.........    0
Free memory.........   52
Program memory......  204


00 00 00 00 00 00 00 00 00 00 00 00 00 00 00 00  ................
00 00 00 00 00 00 00 00 00 00 00 00 00 00 00 00  ................
00 00 00 00 00 00 00 00 00 00 00 00 00 00 00 00  ................
00 00 00 00 F0 BA B5 9D EC 2A 2C 09 88 35 56 F1  ....PPPPPPPPPPPP
F7 57 F1 7A CD 8B 64 69 A5 6E D6 21 10 4F B2 B4  PPPPPPPPPPPPPPPP
F6 96 7C 88 57 54 5A 96 00 02 3E 64 5A 45 57 46  PPPPPPPPPPPPPPPP
B6 B4 5A 03 BA B2 30 BB 5C C5 89 D5 1A D0 95 0F  PPPPPPPPPPPPPPPP
E9 CD 2E 5F 8F 38 B1 5A C3 0C 72 AA B0 CF 20 5B  PPPPPPPPPPPPPPPP
01 22 6A 4E 46 D4 9C D0 A8 39 2B 7A 4E D0 A6 0A  PPPPPPPPPPPPPPPP
8D 9A B3 A2 E6 04 7D AA D0 A8 39 2B 5A 4E D0 A8  PPPPPPPPPPPPPPPP
2A 64 AC 02 05 76 15 27 51 A5 F8 92 44 4D AC 43  PPPPPPPPPPPPPPPP
A2 22 96 DF A1 A2 2A BA 04 FA 67 6C C1 91 5E B1  PPPPPPPPPPPPPPPP
BC 3E 08 8F 14 55 85 29 AA E2 1C 55 8B 02 BA 3E  PPPPPPPPPPPPPPPP
27 1B 66 C8 88 2B 4A 4E D0 AC 5A D1 71 82 6E D5  PPPPPPPPPPPPPPPP
8A 8A 13 B4 A9 56 34 9C A0 4F B5 A2 E0 04 8D AA  PPPPPPPPPPPPPPPP
42 C6 8A B5 D5 F4 14 85 8A D4 0A 84 8D 95 C0 34  PPPPPPPPPPPPPPPP
```

From the memory dump, you can see that 52 bytes are available for further enhancements. In spite of the extensive memory requirements for the CRC calculation, there is still enough room for the implementation of the functions.

Security in Data Transmission by CRC. A correct data transmission in the BASIC Stamp network will be secured by a CRC. Because this method gives a high level of security, it is used in numerous communication protocols. To help you implement this method in other applications, here is a description of the CRC. Without too much math, we can explain the difference between a CRC and a usual check sum.

To calculate the usual check sum, all concerned bytes are simply added. For a check sum in byte format, the result is equal to the sum of all bytes modulo

256. Table 7.4 shows the calculation of such a check sum for three series of three bytes each. In the left column the resulting check sum for the bytes AA, 55, and C0 is BF. In the middle column only one bit in the first byte was changed. The result is immediately evident. The byte series in the right column, however, shows that an even number of bit errors can compensate each other. Such an error is not detectable by the usual check sum.

To calculate a CRC, another method is used. The bits of the data bytes are thought of as coefficients of a (possibly) very long polynom. This polynom will be divided by another polynom; the result is the CRC. Clearly the second polynom is the key to security. The mathematical theory delivers some possibilities:

CRC-12: $x_{12} + x_{11} + x_3 + x_2 + x + 1$

CRC-16: $x_{16} + x_{15} + x_2 + 1$

CRC-CCITT: $x_{16} + x_{12} + x_2 + 1$

CRC-12 is used for 6-bit data. For data in usual byte format, CRC-16 is mainly used in the U.S.A., and CRC-CCITT is mainly used in Europe.

The definition part in the last listing contains the following line:

```
symbol CRCPOLY = $1021          ' CRC-CCITT
```

The CRC polynom used is fixed to the value 1021_H ($0001\ 0000\ 0010\ 0001_B$). If a "1" stands in the binary equivalent, then this position also exists in the polynom.

Due to the limited resources of the BASIC Stamp for CRC calculation, a direct procedure must be used. Table-oriented procedures need less calculation time than direct procedures, and are preferred in other hardware environments. The subroutine calc_crc demonstrates the calculation of the CRC with a direct procedure:

```
calc_crc:
            CRC = Byte*256 ^ CRC
            for i=0 to 7
                if Bit15 = 0 then shift_only
                CRC = CRC * 2 ^ CRCPOLY
                goto nxt
shift_only:     CRC = CRC * 2
nxt:        next
            return
```

Each new byte for CRC calculation is moved into the Hi byte and exor-ed with the last CRC. Next, each bit of this byte will be left-shifted and when this bit is not equal to zero, an exor operation with the polynom follows. The resulting CRC can perform the next step of CRC calculation.

Table 7.4
Examples of check sum calculation

1. Byte	AA (1010 1010)	AB (1010 1011)	AB (1010 1011)
2. Byte	55 (0101 0101)	55 (0101 0101)	55 (0101 0101)
3. Byte	C0 (1100 0000)	C0 (1100 0000)	BF (1011 1111)
Sum (2 Byte)	1BF	1C0	1BF
Sum mod 256	BF	C0	BF

After this explanation of the basics of CRC, we come back to the BASIC Stamp network and the three-byte message. Receiving the three command bytes from the master, the subroutine will calculate the 16-bit CRC and attach this number to the message. The result is a transmission of five bytes from the master to any slave.

The slave now receives five bytes and builds a check sum over these. In the case of an errorless transmission the result must be equal to zero. When the slave sends its answer back to the master, then the master tests the received five bytes. The slaves build a three-bytes answer and attach the CRC bytes. The master tests all five received bytes and accepts this message only if the CRC is zero.

Controlling the BASIC Stamp Network by PC. In this example, the PC serves as network master. After the program starts, the Windows user interface appears as shown in Figure 7.17.

In the top third of the open window, we can see the pullup menus for changing several parameters and the input fields for the three command bytes from the master to the slaves. In the approximate middle of the window is a button for starting the data transfer to the BASIC Stamp network, using a mouse click or the Enter key.

In the bottom third of the window we find four displays, for all data sent to and received by the slaves, and several status messages. This display can help you to rapidly understand all of the communication in this small network.

The network master program can serve the COM1: or the COM2: interfaces. Figure 7.18 shows the details for changing the serial port. The communication parameters are fixed to 2400,N,8,1. The serial port is changed in the pullup menu. The default is COM2: because COM1: is often used by the mouse.

If the master sends a message with an address that doesn't exist in the network, no slave will answer. A timeout reduces the waiting time for reactions from the slaves. The default is a timeout of one second. Timeouts between one and five seconds are possible, but parts of a second are not possible. See Figure 7.19 for the Timeout Input window.

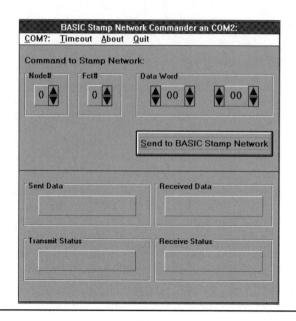

Figure 7.17
Opening window

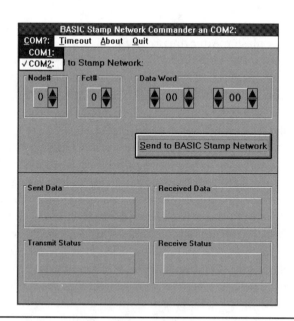

Figure 7.18
Change between COM1: and COM2:, and vice versa

BASIC Stamp

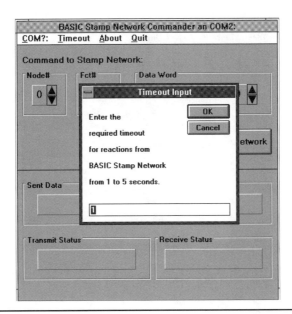

Figure 7.19
Input of timeout

The About window, shown in Figure 7.20, indicates the program version and an E-mail address for feedback. Please pay attention: This E-mail address is invalid.

To stop the program, click the Quit entry with the mouse as usual.

To send a message to the network all boxes in the upper part of the window must be filled in. Figure 7.21 shows the proper entries to send the data word 3456_H to function 2 of node 1.

Clicking the button in the middle of the window sends the data message, with the CRC $DE61_H$ attached, to the network. The master gets no answer, and after the specified time it goes into timeout. This behavior is okay in this case, because no node with the address 1 is in the network.

If the addressed node exists in the network, the whole thing looks totally different. In Figure 7.22, function 0 in node 7 is activated. From the listing of the slave program, we know that a call to function 0 of node 7 will start an A/D conversion. The returned data word is identical to the result of the A/D conversion (here $007F).

Function 1 in node 7 gives back the value of the calibration constant scale. This value was defined as 93 (=$5D). Figure 7.23 shows that the value was read correctly.

Function 2 of node 7 is an example of a general calculation. The data word received by the slave is multiplied by a value of two. The result will be sent back

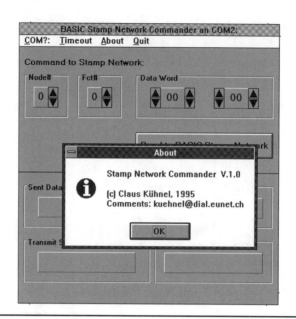

Figure 7.20
About window

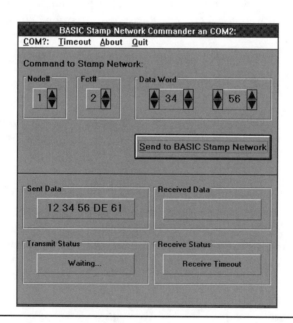

Figure 7.21
Not-answered message to unknown node

BASIC Stamp

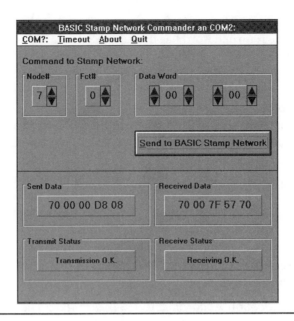

Figure 7.22
Activate function 0 of node 7

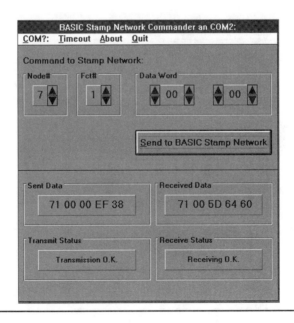

Figure 7.23
Activate function 1 of node 7

to the master. Figure 7.24 shows both messages. The original data word had the value $1234. The returned word has the value $2468.

Calling an unimplemented function, such as function 3 for node 7, gives the value $C0 back in the command byte. Now the master knows that this function is not implemented in this node. The status field displays the error message, as Figure 7.25 shows.

Until now, all data transmissions have been without errors. Figure 7.26 shows that a data transmission had an error, so the recalculation of the CRC must have an error too.

The message sent back to the master signals this error by a set MSB in the command byte. The failed CRC is sent back as data word. The messages in the status fields show the direction in which the error occurred. In Figure 7.26, the error was in the data transmission from master to slave.

7.2.2 Communication Between BASIC Stamps Over Longer Distances

Wired Communication According to RS422/RS485 The communication between BASIC Stamps or the BASIC Stamp and a PC discussed in the last section was based on RS232, and is suitable only for short distances. For a wired communication over longer distances the electrical conditions of the connections are

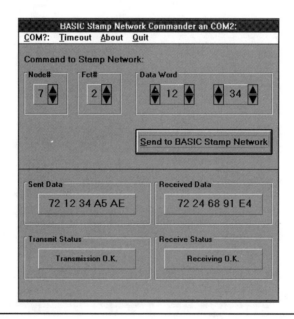

Figure 7.24
Activate function 2 of node 7

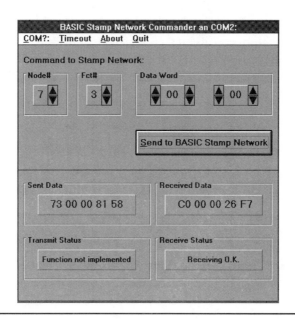

Figure 7.25
Activate a not-defined function in node 7

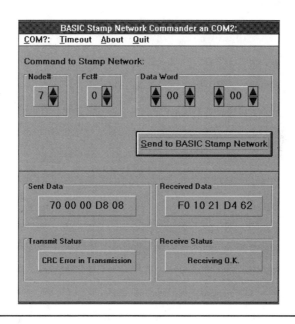

Figure 7.26
Error in data transmission

more and more important. Twisted-pair connections with correctly-terminated lines can serve as an electrical connection between the transmitter and receiver.

RS422 and RS485 are symmetrical interfaces for transmission lines that enhance the features of the traditional current loop and overcome the limits of RS232. They have the following features:

- Maximum data rates of 10 MBit per second.
- A maximum electrical connection length of 1200 m.
- Symmetrical transmission to lessen the chance of electrical disturbances.

Not all features of RS422 or RS485 are relevant for the BASIC Stamps, but if data can be transmitted via twisted-pair wires over distances of some hundred meters, then we already have great results.

RS422A is the specific version of RS422 and makes possible a one-way (simplex) data transmission. There is one transmitter, but there can be more than one receiver. RS485 is an enhancement of the RS422 interface; it makes possible bi-directional (half duplex) data transmission via twisted-pair connections.

Some features of the RS422A and RS485 interfaces are shown in Table 7.5 and Figure 7.27. In the upper half of Figure 7.27, one driver feeds the symmetrical line, terminated with a resistor of 100 Ω. According to the RS422A standard, up to ten receivers may be connected to the symmetrical line. In the lower half of the Figure, we have a complete bi-directional two-wire bus system. Up to 32 transceivers (that means transmitter and receiver) can be placed around the bus.

Nearly each manufacturer of semiconductors offers its own transceiver devices. Figure 7.28 shows the logic diagram of SN75176A from Texas Instruments, as a random example. On the right, the markers A and B characterize both symmetrical bus lines. The driver with its symmetrical output is connected to the symmetrical input of the receiver in the internal transceiver device. The communication, according to RS485, works half-duplex. That means that either the transmitter or the receiver is active and the control lines DE (Driver

Table 7.5
Some features of the RS422A and RS485 interfaces

	RS422A	RS485
Common-mode input voltage	−7 V . . . +7 V	−7 V . . . +12 V
Receiver input resistance	4 kΩ minimum	12 kΩ minimum
Driver load resistance	100 Ω	60 Ω
Short-circuit output current	150 mA to GND	150 mA to GND
		250 mA to −7 V or +12 V

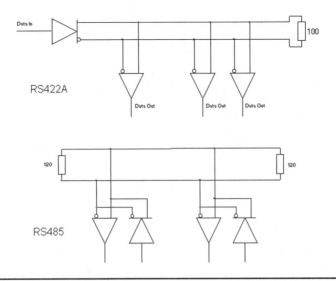

Figure 7.27
Interfaces according to RS422A and RS485

Enable) and /RE (Receiver Enable) can be connected together. Hint: /RE means an Lo-active input RE.

For the BASIC Stamp there are only very small changes, compared with the serial interface of RS232. The connected enable inputs will be controlled by an additional I/O pin of the BASIC Stamp. Before a SEROUT instruction can be executed, the enable input must be switched to Hi; after the execution of SEROUT, it must be immediately switched to Lo again to reactivate the receiver circuit.

[BS1]—Communication via Radio Frequency For longer distances a two-wire connection can be difficult to handle, so wireless techniques are

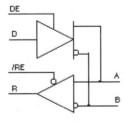

Figure 7.28
Logic of SN75176A

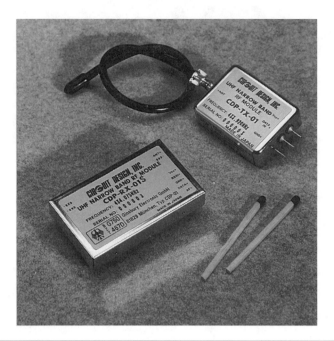

Figure 7.29
RF Modules for data transmission

preferred. The market provides some solutions for wireless data transmission. We have to look in the low-end of devices for wireless data transmission, because the BASIC Stamp microcontroller is on the lower end.

Circuit Design Inc. (Japan) offers RF modules for data transmission that are well-suited for BASIC Stamp applications. Figure 7.29 shows such RF modules for data transmission in simplex mode. The transmitter and receiver of the RF module are TTL-compatible and will take over all functions required for transmission of pulses or data. Both modules are entirely metallic encapsulated.

The FM narrow-band modulation used gives better results than the usual AM or FM wide-band modulation. The resulting range can be measured.

An integrated voltage control in the transmitter and receiver provides stable transmission and reception, even when the voltage supply changes. The current consumption is very low so that the equipment is suitable for battery operation.

Figure 7.30 shows how simply a wireless data transmission based on these RF modules can be built. The TTL-compatible data input of the transmitter is controlled by an RS232 data output (Pin0) from the first BASIC Stamp I. The supply voltage of the transmitter module is defined between 5.2 V and 10 V DC.

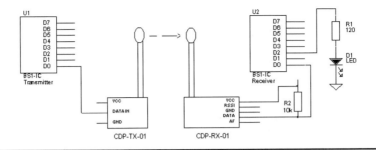

Figure 7.30
Wireless data transmission with RF modules

Deviating from this specification, we feed the transmitter side with one +5 V DC power supply.

The data output of the receiver is designed as an open-collector. A pullup resistor of about 10 kΩ must be connected to V_{CC}. For the receiver side, a supply voltage of +5 V DC is quite sufficient.

With an NF bandwidth from 100 Hz to 3000 Hz, a baud rate between 300 Baud and 2400 Baud will work for communication.

The next two listings show a simple example of switching an LED over a wireless communication link. Because this RF module works in simplex mode only, no return message is possible.

The requirements for a half-duplex data transmission are the same as for the RS485 interface described. RF modules for a bi-directional (half-duplex) data transmission have a switch for changing the data direction similar to the switch on the transceiver device SN75176A. Usual RF modules for half-duplex communication have a considerable time delay for switching the data direction. The whole handling for half-duplex communication will be a little more difficult then the simple one-way communication described here.

```
' -----[ Title ]------------------------------------------------
'
' File......  RADIOTRA.BAS
' Purpose...  Airborn Control
' Author....  Claus Kühnel
' Started...  03.10.94
' Updated...


' -----[ Program Description ]-----------------------------------
'
' A radio transmitter sends RS232 commands to an radio receiver for
' control purposes.
```

```
' -----[ Revision History ]-----------------------------------
'
'   3.10.94: Version 1.0

' -----[ Constants ]-----------------------------------------
'
        symbol TxD = 0                ' RS232 output
        symbol baud = N1200           ' 1200 Baud

' -----[ Variables ]-----------------------------------------
'
        symbol command = b0           ' command byte for the receiver

' -----[ Initialization ]------------------------------------
'

' -----[ Main Code ]-----------------------------------------
'
start: command = "1"                  ' send command byte
       gosub send

       command = "0"                  ' send another command byte
       gosub send

       goto start                     ' repeat endless

' -----[ Subroutines ]---------------------------------------
'
send:  serout TxD, baud, (command)    ' send the command to the slave
       pause 1000                     ' wait a little bit
       return
```

```
' -----[ Title ]---------------------------------------------
'
' File......  RADIOREC.BAS
' Purpose...  Airborn Control
' Author....  Claus Kühnel
' Started...  30.09.94
' Updated...

' -----[ Program Description ]-------------------------------
'
' A radio receiver gets commands from a transmitter to control
' peripheral functions (here switching an I/O line for
' demonstration).
```

```
' -----[ Revision History ]--------------------------------------
'
'   3.10.94: Version 1.0


' -----[ Constants ]---------------------------------------------
'
        symbol RxD = 0              ' RS232 input
        symbol OUT = 2              ' pin for connecting a LED
        symbol baud = N1200        ' 1200 Baud


' -----[ Variables ]---------------------------------------------
'
        symbol command = b0        ' command byte from the master


' -----[ Initialization ]----------------------------------------
'


' -----[ Main Code ]---------------------------------------------
'
start: serin RxD, baud, command    ' look for a command from the
                                     master
       if command = "1" then PinHi ' set pin Hi
       if command = "0" then PinLo ' set pin Lo
       goto start                  ' ignore other commands


PinLo:low OUT                      ' set pin Lo
       goto start


PinHi:high OUT                     ' set pin Hi
       goto start


' -----[ Subroutines ]-------------------------------------------
'
```

7.2.3 [BS2]—Remote Alarm Message via Modem

In recent years, phone lines have gone from only making connections between human callers to transmitting data in several ways. The link between a phone line and a serial port of the PC or other devices is the **modem.** The connection between the modem and the controlling device is standardized for byte transmission according to RS232.

   This connection is the only one. So there are two different messages to transmit:

- The first is the initializing of the modem for the transmission, according to the possibilities of the line and the parameters of the remote modem. Included here are dialing the remote station and closing the connection after successful transmission, and generating messages of success or fault. This is the command mode, reached after connecting the modem with power.

    The control data required to work in this mode is standardized by HAYES statements. These statements for modems are similar to a programming language with its own syntax and special words. In Appendix A, the "Hayes Modem Instruction Set" with the most important statements is shown. How to use any other language: There are many words and express combinations. But you need and should use only the important ones. All statements start with "AT." (There are only two important expressions.) The most often used initializing contracts for communication are stored in the standard command "ATZ."

- After initialization, including the test of connection in both directions, the modem goes to its **transparent mode** automatically. This means that both modems are successfully connected and ready for data transmission. Transparent means that, like a mirror, all data output from the transmitter's RS232 plug conform with the RS232 plug of the receiver's station. If there is no modem in the transmission chain, the connection is similar to a short-circuited RS232 connection with no phone line.

Keep in mind that after transmitting or receiving data, the transparent mode must be closed to hang up the phone line in command mode. Then the modem is like an unused phone, a passive device. If you want a new connection, you must initialize as shown above.

That's all. You need to understand how to connect and program, as explained below. But first, the following is important.

The modem and one of the BASIC Stamps are in accord in more than one feature. They do not require much power. And, unlike a PC, there are no mechanical components such as motors for cooling and driving hard disks. If you working with only a modem and a BASIC Stamp, there is no noise. Both components are cheap, and last but not least, they have a RS232 port for data transmission. So it is advantageous for some purposes to couple these components to work silently, full time in the background.

In the following example, we want to give an alarm message to a remote station automatically. The remote location needs a working PC with connected modem. We demonstrate the transmission only with BASIC Stamp II.

In the same way, it is possible to use the BASIC Stamp for processing received data. For example, a closed contact may start the alarm message. We used a key on Pin14, shown before. The event for an alarm may be chosen by the user. The event may be a combination of input lines and their conditions, several sensors, keys, or analog-to-digital converters.

Figure 7.31 shows the hardware connections. The TxD wire and GND are needed for data connection between the modem and BASIC Stamp. But to simulate a full hardware protocol we added one further wire. The successful connection after dialing is tested by the BASIC Stamp with the signal "DCD." That means with the initializing HAYES Command, "&C1" (default set) provides "DCD follows the carrier."

The carrier detection is a result of successful dialing and an actual usable connection. The DCD output is a positive-going voltage, in keeping with the RS232 level "Space" for an active DCD. With two diodes, we clamp the signal to values safe for the input P7 of our BASIC Stamp II. So, modem and BASIC Stamp are connected with the three wires TxD, DCD, and GND.

But you cannot add two additional wires on the modem side to complete the hardware protocol. Pulling up the line DTR to VCC = 5 V simulates the terminal output "Data Terminal Ready." The modem permits data transmission to a terminal—We use the BASIC Stamp.

We complete the connections with wiring CTS to RTS. CTS signifies "Clear To Send" as an output and provides permanently-set "Ready To Send" to the modem input, like a hardware self-initializing of the modem itself.

The program used (File ALARM.BS2) is shown in the following listing:

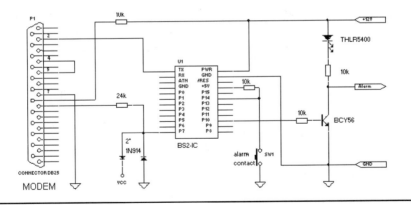

Figure 7.31
Alarm device with modem connection

```
' -----[ Title ]----------------------------------------------------
'
' File......alarm.BS2
' Purpose...alarm-message by phone with BS2 and modem
' Author....Klaus Zahnert
' Started...95.12.10
' Updated...


' -----[ Program Description ]----------------------------------
' alarm or any other importand event  is marked with one input-pin
' going  low. In this case modem is initialised, makes connection
' with dial and gives alarm to remote station. The connection
' between stations is tested by DCD-signal. Alternative alarm is
' giving for connection is  failured after dial.


' -----[ Revision History ]------------------------------------
'
'


' -----[ Constants ]---------------------------------------------
'
          txdpin     con 16           'standard rs232 BSTAMP2
          dcdpin     con 07           'connected with modem - DCD-output
          sirenpin   con 15           'output to siren/LED
          alarm      con 4711         'typical number of alarm-message
          bdmd       con 84+$4000     '9600 Baud, 8n1
          delay      con 1000         'wait for branch


' -----[ Variables ]---------------------------------------------
'
          contact    var   IN14       'Key to ground with pullup to Vcc


' -----[ Initialization ]----------------------------------------
'
          DIRS = $8000         ' output sirenpin
          low sirenpin         ' no siren


' -----[ Main Code ]---------------------------------------------
'


start:   if contact = 0  then message        'alarm if keypressed
         goto start
          'debug "     "                       'only for test
```

```
message: serout txdpin,bdmd, ["ATZ",10,13]        'modem standard init
        pause 2000
        serout txdpin,bdmd, ["ATDT 1234567",10,13]        'dial remote station
        pause 20000
        serout txdpin\dcdpin,bdmd,delay,siren,["ALARM = ",dec alarm,10,13]
                                                   'out alarm-message


        pause 3000
        serout txdpin,bdmd,10,["+++"]               'ret. to command-mode
        pause 3000
        serout txdpin,bdmd, ["ATH",10,13]           'hang up


        goto start

siren: high 15                                    'siren/LED on
        goto start              'try again alarm-message if contact = low


end


' -----[ Subroutines ]---------------------------------------------
'no
```

You must set up your own alarm number or any other ASCII-based message. Also you need to set up the phone number of the remote station with all digits needed.

The program uses the command serout for all phases of data transmission to the modem. The connection is started after initializing and dialing—or not, if the remote station is busy or in failure.

So we test the line DCD with the enhanced statement SERIN of BASIC Stamp II. The statement includes the optional branch to mark with the test of a pin. The program jumps to the mark siren, if the connection is not successful, i.e. no carrier arrives from the remote station. A further SEROUT of the alarm message makes no sense. So another signal, for example a siren, will be started in place of the actual alarm. Hopefully, it will help, depending on the situation!

You will notice the splitting commands "+++" (in command mode) and "ATH." Often they are coupled with one serout instruction.

We find on other microcontrollers (for example on the assembler programmed M68HC11), that this delay is often needed for working. We successfully use the longer time for interpreting single statements for serout.

In good modem manuals or descriptions of HAYES commands, you will find special time conditions for the "+++" command. These are needed for the modem to distinguish between a single "+" to transmit like any other byte, or to translate a string of the following two "+"s to the inner modem command, "go return from transparent mode to command mode." For that purpose, a time is stored in modem register S12. The default value is 50 for a time 50/50 sec = 1 sec. Only if the time = 1 sec before and after, giving "+++" no bytes are sent to the modem, and the time between the "+"s is smaller than that value (7 sec), then the given string is recognized as a command and not as single bytes.

Here is some advice for a careful installation of this program. Before connecting with a defined remote station in this way, you should take steps to avoid trouble for any mislinked stations and for you. These steps are

- Transfer the program, shown above. Use **ALT-M** to test "compile o.k."
- Connect BASIC Stamp II and run with `debug` each `serout`.
- Connect modem, as shown in Figure 7.31.
- Insert your *own* phone number and your special alarm message. For this step only simulate "Carrier Detect" with an external voltage of +12 V instead of connecting with the modem. Run with it to get "Busy." The speaker of the modem and the display or LED's control behavior shows success or mistakes.
- If you were successful, replace the number for starting the connection with the remote station. This station must have started its own terminal program before.

7.2.4 [BS1]—Connecting a Keyboard

A keyboard is one standard input device for computers or other computer-controlled equipment. Pressing local distributed contacts also serves as an input medium for control units.

The BASIC Stamp with its limited I/O pins can only utilize a small number of keys without additional circuitry. If the keys are arranged in a matrix, the maximum number of keys is equal to the product of the row and column lines. BASIC Stamp I has eight I/O lines. If one pin is reserved for a serial output, then seven I/O pins are available for utilizing the keyboard. For the BASIC Stamp II with its sixteen free I/O pins, the situation is not such a problem. Table 7.6 shows the available number of keys for both BASIC Stamps.

To find a pressed key in a matrix with columns and rows, the I/O lines in one direction (especially the rows) are set to Hi one after the other. When a row is Hi all lines are queried, one after the other. By this arrangement of keys and that kind of query, the pressed key is noted and can be evaluated by a query program.

In the keyboard controller shown in Figure 7.32, a BASIC Stamp I queries a 3 × 4 key matrix and outputs the key code to a serial output. Besides the ten figures, two characters ("*" and "#") exist. These characters can serve as control signals, for example.

The whole listing for a simple keyboard controller is given with the listing TAST.BAS:

Table 7.6
Available number of keys

	BASIC Stamp I		BASIC Stamp II	
Available I/O Pins	8	7	16	15
Number of Keys	16	12	64	56

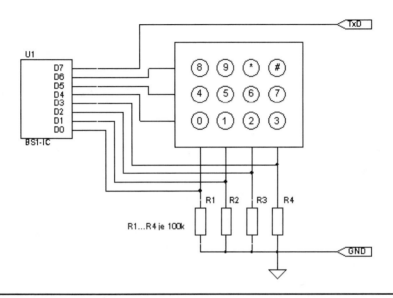

Figure 7.32
BASIC Stamp as keyboard controller

```
' -----[ Title ]-----------------------------------------------

'
' File......TAST.BAS
' Purpose...keying number 0....11
' Author....Klaus Zahnert
' Started...
' Updated...


' -----[ Program Description ]----------------------------------
' four pins input and three pins outp. are crossed in 12 key -
' matrix. Inputs are connected with resist. 100 k to ground. Number
' of pressed key is stored in var. "Zahl" for serial output in
' subroutine outp.
```

```
'-----[ Revision History ]---------------------------------------------
'        REV 0


'-----[ Constants ]-----------------------------------------------------
'
        symbol baud        = n2400
        symbol txd         = 7


'-----[ Variables ]-----------------------------------------------------
'
        symbol connect     = pins    'pin 0..3 input: column
                                     'pin 4..6 output: row
                                     'pin 7     serial output RS232
        symbol zeile       = b2      'count row
        symbol spalte      = b3      'count column
        symbol zahl        = b4      'result of keying
        symbol mark        = b5      'column binary value
        symbol sp          = b6      'column position


'-----[ Initialization ]------------------------------------------------
        dirs = %01110000             ' 1 = output
'                                      0 = input
'-----[ Main Code ]-----------------------------------------------------
'
loop:   zahl = 0
        mark = 0
        for zeile = 4 to 6
          high  zeile                'each col. high with asking all rows
              for spalte = 0 to 3
                  mark = connect & $0f    'mask row
               if mark > 0  then val      'if keypressed, one column
                                          'is going high

            next spalte
          low   zeile
          zahl = zahl + 4
        next zeile
        goto loop


val:    lookdown mark,(8,4,2,1),sp       'binary val. to position col.
        zahl = zahl + sp                 'calculate row / column
        gosub outp                       'subroutine output


lp1:    mark = connect & $0f             'waiting loop while
        if mark > 0 then lp1             'key is pressed
        goto loop
```

```
'  -----[ Subroutines ]------------------------------------------------
'
outp: serout txd,baud,("Input number = ",#Zahl,13,10) 'serial output
      return
```

The lines for querying the column are open CMOS inputs and have to get pulldown resistors. Without these pulldown resistors, the CMOS inputs could have undefined levels if they are not switched to Hi by a pressed key. These undefined levels would give faulty key codes on the serial output.

Connecting a serial input to the serial output of our keyboard controller, we can interpret the signals resulting from pressed keys. The speed of reaction is quick enough. While the key is being pressed, the waiting loop lp1 prevents another query.

7.2.5 Control of Digital Displays

In connection with the BASIC Stamps, all display units without complex interfaces are suitable for display of alpha-numerical or numerical information. The number of I/O lines and the memory required for the implementation of complex protocols are both restricted.

Some display units reduce the required control complexity for the control unit by integrating their own controllers.

[BS1]—Driving an LCD   Liquid crystal displays (LCDs) have very small current consumption and are very suitable for working with the BASIC Stamp. The LCD controller HD44780 from Hitachi is regarded as an industrial standard for control of alpha-numerical LCDs. LCDs of different sizes with an integrated HD44780 controller are offered by numerous companies.

In the following program examples, an LCD with four rows and sixteen characters, such as LM041L from Hitachi, was used. It is possible to use each LCD with an integrated HD44780. Figure 7.33 shows the display unit LM041L.

On the top of the display are 14 pins, whose meaning is explained in Table 7.7. All electronic devices are soldered on the back side and therefore cannot be seen in Figure 7.33.

With 14 I/O pins of this and other LCDs, one could suppose that the BASIC Stamps are overtaxed. However, the data bus width can be reduced to four bits and then the relation of the I/O lines looks better. Figure 7.34 shows the complete hardware for an LCD driven by a BASIC Stamp I. From the original eight-bit wide data bus, only the four higher bits are used for the four-bit data bus mode.

Figure 7.34 shows clearly that only six connections are required between BASIC Stamp I and the LCD module. All data were transferred to the LCD via four data lines. Any change from Hi to Lo on line E latches the data into the LCD controller. The level on line RS differs between commands and data.

Figure 7.33
LCD module LM041L (Hitachi)

Table 7.7
LM041L - Electrical connections

Pin	Meaning	Level	Function
1	V_{SS}	GND	Ground
2	V_{DD}	+5 V	Supply voltage
3	Vo	0 ... +5 V	Contrast control
4	RS	H/L	L: Input Command
			H: Input Data
5	R/W	H/L	L: Read data from display
			H: Write data to display
6	E	H → L	Enable
7	DB0	H/L	
8	DB1	H/L	
9	DB2	H/L	
10	DB3	H/L	Data bus
11	DB4	H/L	
12	DB5	H/L	
13	DB6	H/L	
14	DB7	H/L	

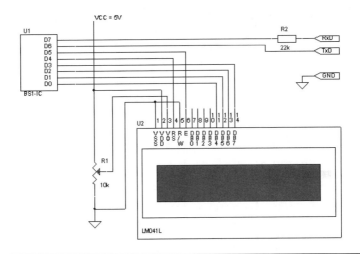

Figure 7.34
LCD LM041L driven by BASIC Stamp I

Before a first look at the program examples, some information is required about the LCD controller HD44780. Only enough information as is needed for understanding and adapting the following program examples will be given. Detailed information is contained in the voluminous data books of the many manufacturers of LCDs and LCD controllers.

The LCD controller HD44780 has two internal 8-bit registers accessable by the BASIC Stamp. The instruction register (IR) stores the commands (RS = 0), while the data register (DR) temporarily stores data (RS = 1) for the data display RAM (DD RAM) or the character generator RAM (CG RAM).

The DD RAM contains 80 bytes, so the maximum display size is four rows with 20 characters. Table 7.8 provides an overall view of the addressing locations on the LCD and the address of the RAM cell for a 4 × 16-LCD (LM041L, EA7164-N, EA8164-ANLED, EAVK-2416, and others).

From Table 7.8 you can see that not all memory cells serve as display memory. The unused memory cells can be used as external memory. To reduce the required I/O lines for the interface, the R//W line is connected to GND. Under this condition, a read operation is impossible. If you want to use these memory locations as additional memory, you need an additional pin from the BASIC Stamp.

For programming the LCD, Table 7.9 shows an extract of the command set for the LCD controller HD44780. The read operations prevented by the deactivated line R//W are not explained here.

Table 7.10 shows some explanations to the command set.

With this description, the program examples can be interpreted and adapted to actual applications. The next two programs were developed by Jon Williams

Table 7.8
Relationship between locations on the LCD and addresses in DD RAM

DDRAM	1	2	3	4	5	6	7	8	9	10	11	12	13	14	15	16
1. Row	00	01	02	03	04	05	06	07	08	09	0A	0B	0C	0D	0E	0F
2. Row	40	41	42	43	44	45	46	47	48	49	4A	4B	4C	4D	4E	4F
3. Row	10	11	12	13	14	15	16	17	18	19	1A	1B	1C	1D	1E	1F
4. Row	50	51	52	53	54	55	56	57	58	59	5A	5B	5C	5D	5E	5F

Table 7.9
Instruction coding for LCD controller HD44780

Instruction	RS	DB7	DB6	DB5	DB4	DB3	DB2	DB1	DB0	Description
Clear Display	0	0	0	0	0	0	0	0	1	Clear the display and set cursor in home position
Cursor At Home	0	0	0	0	0	0	0	1	X	Set cursor in home position (left top = DD RAM Addr = 00H)
Set Entry Mode	0	0	0	0	0	0	1	I/D	S	Set direction for cursor movement and shiftmode
Display On/Off	0	0	0	0	0	1	D	C	B	Look to the explanations to D, C, and B
Cursor/Display Shift	0	0	0	0	1	S/C	R/L	X	X	Look to the explanations to S/C and R/L
Function Set	0	0	0	1	DL	N	F	X	X	Look to the explanations to DL, N, and F
Set CG RAM Addr	0	0	1				ACG			Set CG RAM Address
Set DD RAM Addr	0	1				ADD				Set DD RAM Address
Data Write	1				Data					Write data in DD RAM or CG RAM

Table 7.10
Explanations to the command set of LCD controller HD44780

I/D	DD RAM or CG RAM address will be incremented (I/D = 1) or decremented (I/D = 0) after writing one character into the RAM.
S	Shifts the whole display contents right (S = 1) or left (S = 0). It seems that the cursor stays at its old position (calculator). For I/D = 1 the display will be shifted to the left, for I/D = 0 to the right.
D	Display on (D = 1) or off (D = 0). Data in DD RAM stay unchanged.
C	Cursor is displayed (C = 1) or not (D = 0).
B	Cursor blinks (B = 1) or not (B = 0).
S/C	Shifts display contents (S/C = 1) or cursor (S/C = 0) according to R/L one position.
R/L	Shifts to the right (R/L = 1) or left (R/L = 0) without changing data in DD RAM.
DL	Data lenght 8 Bit (DL = 1) or 4 Bit (DL = 0).
N	Number of rows in the display - one (N = 0) - more (N = 1).
F	Font - 5 × 7 Dots (F = 0) - 5 × 10 Dots (F = 1).
X	Don't care.

for an LCD with 16 characters (Optrex DMC-16106). Both programs are documented very well, so only a few explanations should be needed. The first listing shows a program demonstrating some display functions (File LCDDEMO1.BAS).

```
' -----[ Title ]----------------------------------------------
'
' File...... LCDDEMO1.BAS
' Purpose... Stamp -> LCD (4-bit interface)
' Author.... Jon Williams
' Started... 16 July 1994
' Updated... 16 July 1994


' -----[ Program Description ]--------------------------------
'
' This program demonstrates the various standard features of an LCD
' display that uses the Hitatchi HD44780 controller. The LCD used
' to test this program was the Optrex DMC-16106 (16x1).
'
' LCD Connections:
'
' LCD            (Function)         Stamp
' --------------------            -----
' pin 1          Vss                gnd
' pin 2          Vdd                +5
```

166   *BASIC Stamp*

```
' pin 3          Vo              gnd
' pin 4          RS              pin 4
' pin 5          R/W             gnd
' pin 6          E               pin 5
' pin 7          DB0             gnd
' pin 8          DB1             gnd
' pin 9          DB2             gnd
' pin 10         DB3             gnd
' pin 11         DB4             pin 0
' pin 12         DB5             pin 1
' pin 13         DB6             pin 2
' pin 14         DB7             pin 4


' -----[ Revision History ]-----------------------------------------
'
' 07-16-94 : Version 1.0 - compilation of code from last 3 months


' -----[ Constants ]------------------------------------------------
'
SYMBOL  E       = 5                     ' LCD enable pin (1 = enabled)
SYMBOL  RS      = 4                     ' Register Select (1 = char)

' LCD control characters
'
SYMBOL  ClrLCD  = $01                   ' clear the LCD
SYMBOL  CrsrHm  = $02                   ' move cursor to home position
SYMBOL  CrsrLf  = $10                   ' move cursor left
SYMBOL  CrsrRt  = $14                   ' move cursor right
SYMBOL  DispLf  = $18                   ' shift displayed chars left
SYMBOL  DispRt  = $1C                   ' shift displayed chars right


' -----[ Variables ]------------------------------------------------
'
SYMBOL  outp    = B0                    ' output workspace
SYMBOL  char    = B1                    ' char sent to LCD
SYMBOL  index   = B2                    ' loop counter

' -----[ Initialization ]-------------------------------------------
'
        BSAVE                           ' save EEPROM image file
        EEPROM ("THE BASIC STAMP!")     ' preload EEPROM

        Pins = %00000000                ' clear the pins
        Dirs = %00111111                ' set 0-5 as outputs
        PAUSE 20                        ' let the LCD settle
```

BASIC Stamp Applications   167

```
' Initialize the LCD (Hitatchi HD44780 controller)
'
LCDini: Pins = %00000011              ' 8-bit mode
        PULSOUT E, 1
        PAUSE 5
        PULSOUT E, 1
        PULSOUT E, 1
        Pins = %00000010              ' 4-bit mode
        PULSOUT E, 1
        char = %00001100              ' disp on, crsr off, blink off
        GOSUB WrLCD
        char = %00000110              ' inc crsr, no disp shift
        GOSUB WrLCD
        char = %00000001              ' clear LCD
        GOSUB WrLCD
        HIGH RS                       ' LCD to character mode


' -----[ Main Code ]-------------------------------------------
'
Start:  FOR index = 0 TO 15
          READ index, char           ' get char from EEPROM
          GOSUB WrLCD                 ' write it
          PAUSE 50                    ' delay between chars
        NEXT index
        PAUSE 2000                    ' wait 2 seconds

        char = CrsrHm                 ' move the cursor home
        GOSUB LCDcmd

        char = %00001110              ' cursor on
        GOSUB LCDcmd
        PAUSE 500

        char = CrsrRt
        FOR index = 1 TO 15           ' move the cursor
          GOSUB LCDcmd
          PAUSE 100
        NEXT index

        FOR index = 14 TO 0 STEP -1   ' go backward by moving to
          char = %10000000 | index    '  a specific address
          GOSUB LCDcmd
          PAUSE 100
        NEXT index
```

```
        char = %00001101                    ' cursor off, blink on
        GOSUB LCDcmd
        PAUSE 2000


        char = %00001100                    ' blink off
        GOSUB LCDcmd


        FOR index = 1 TO 10                  ' flash display
          char = char ^ %00000100            ' toggle display bit
          GOSUB LCDcmd
          PAUSE 250
        NEXT index
        PAUSE 1000


        FOR index = 1 TO 16                  ' shift display
          char = DispRt
          GOSUB LCDcmd
          PAUSE 100
        NEXT index
        PAUSE 1000


        FOR index = 1 TO 16                  ' shift display back
          char = DispLf
          GOSUB LCDcmd
          PAUSE 100
        NEXT index
        PAUSE 1000


        char = ClrLCD                        ' clear the LCD
        GOSUB LCDcmd
        PAUSE 500


        GOTO Start                           ' do it all over

' -----[ Subroutines ]----------------------------------------------
'
' Send command to the LCD
'
' Load char with command value, then call
'
'   Clear the LCD............. $01, %00000001
'   Home the cursor........... $02, %00000010
'   Display control........... (see below)
'   Entry mode................ (see below)
```

```
'    Cursor left............... $10, %00010000
'    Cursor right.............. $14, %00010100
'    Scroll display left....... $18, %00011000
'    Scroll display right...... $1C, %00011100
'    Set CG RAM address........        %01aaaaaa (Character Generator)
'    Set DD RAM address........        %1aaaaaaa (Display Data)
'
' Display control byte:
'
'    % 0 0 0 0 1 D C B
'                    | | -- blink character under cursor (1=blink)
'                    | ---- cursor on/off (1=on)
'                    ------ display on/off (1=on)
'
' Entry mode byte:
'
'    % 0 0 0 0 0 1 X S
'                    |--- shift display (S=1), left (X=1), right (X=0)
'                    ---- cursor move: right (X=1), left (X=0)
'
LCDcmd: LOW RS
        GOSUB WrLCD
        HIGH RS
        RETURN


' Write ASCII char to LCD
'
WrLCD:  Pins = Pins & %00010000        ' RS = 1, data bus clear
        outp = char / 16               ' get high nibble
        Pins = Pins | outp             ' output the nibble
        PULSOUT E, 1                   ' strobe the Enable line
        Pins = Pins & %00010000
        outp = char & %00001111        ' get low nibble
        Pins = Pins | outp
        PULSOUT E, 1
        RETURN
```

Section Constants defines the control characters according to Table 7.9. The initialization of the LCD controller starts at the label LCDini. After this initialization, the line RS is set to Hi and all characters following will be moved into DD RAM.

Two other subroutines are responsible for further activities. The subroutine LCDcmd will for a short time switch into command mode (LOW RS) to send one character with the subroutine WrLCD to the LCD controller. Due to the four-Bit data bus, we have to use two steps to send one character—Hi nibble first and Lo nibble second.

In addition to the 192 characters already defined, eight characters (ASCII 0 to ASCII 7) can be defined by the user. The CG RAM address (ACG in Table 7.9) consists of the six bits shown here in Table 7.11:

Table 7.11
Contents of CG RAM

DB5	DB4	DB3	DB2	DB1	DB0
	ASCII Code			Pixel Row	

Eight pixel rows build one character. The bottom pixel row is identical to the cursor line and is free in general.

The following program example LCDDEMO2.BAS defines three user-defined characters well known from many games. In a moving sequence these characters guzzle text from display.

```
' -----[ Title ]-------------------------------------------------
'
' File...... LCDDEMO2.BAS
' Purpose... Stamp -> LCD (4-bit interface) with custom characters
' Author.... Jon Williams
```

Figure 7.35
User defined characters

```
' Started... 16 July 1994
' Updated... 16 July 1994


' -----[ Program Description ]-------------------------------------
'
' This program demonstrates the generation of custom characters for
' an LCD display that uses the Hitatchi HD44780 controller. The LCD
' used to test this program was the Optrex DMC-16106 (16x1).
'
' LCD Connections:
'
' LCD            (Function)            Stamp
' --------------------                 -----
' pin 1          Vss                   gnd
' pin 2          Vdd                   +5
' pin 3          Vo                    gnd
' pin 4          RS                    pin 4
' pin 5          R/W                   gnd
' pin 6          E                     pin 5
' pin 7          DB0                   gnd
' pin 8          DB1                   gnd
' pin 9          DB2                   gnd
' pin 10         DB3                   gnd
' pin 11         DB4                   pin 0
' pin 12         DB5                   pin 1
' pin 13         DB6                   pin 2
' pin 14         DB7                   pin 4


' -----[ Revision History ]---------------------------------------
'
' 07-16-94 : Version 1.0


' -----[ Constants ]----------------------------------------------
'
SYMBOL  E       = 5              ' LCD enable pin (1 = enabled)
SYMBOL  RS      = 4              ' Register Select (1 = char)

' LCD control characters
'
SYMBOL  ClrLCD  = $01            ' clear the LCD
SYMBOL  CrsrHm  = $02            ' move cursor to home position
SYMBOL  CrsrLf  = $10            ' move cursor left
SYMBOL  CrsrRt  = $14            ' move cursor right
SYMBOL  DispLf  = $18            ' shift displayed chars left
SYMBOL  DispRt  = $1C            ' shift displayed chars right
```

```
' -----[ Variables ]-------------------------------------------------
'
SYMBOL   outp    = B0              ' output workspace
SYMBOL   char    = B1              ' char sent to LCD
SYMBOL   index1  = B2              ' loop counter
SYMBOL   index2  = B3              ' loop counter


' -----[ Initialization ]--------------------------------------------
'
       BSAVE                       ' save EEPROM image file
       EEPROM ($0E,$1F,$1C,$18)    ' char 0 top
       EEPROM ($1C,$1F,$0E,$00)    ' char 0 bottom
       EEPROM ($0E,$1F,$1F,$18)    ' char 1 top
       EEPROM ($1F,$1F,$0E,$00)    ' char 1 bottom
       EEPROM ($0E,$1F,$1F,$1F)    ' char 2 top
       EEPROM ($1F,$1F,$0E,$00)    ' char 2 bottom
       EEPROM ("THE BASIC STAMP!") ' display string


       Pins = %00000000           ' clear the pins
       Dirs = %00111111           ' set 0-5 as outputs
       PAUSE 20                    ' let the LCD settle

' Initialize the LCD (Hitatchi HD44780 controller)
'
LCDini: Pins = %00000011           ' 8-bit mode
       PULSOUT E, 1
       PAUSE 5
       PULSOUT E, 1
       PULSOUT E, 1
       Pins = %00000010           ' 4-bit mode
       PULSOUT E, 1
       char = %00001100           ' disp on, crsr off, blink off
       GOSUB WrLCD
       char = %00000110           ' inc crsr, no disp shift
       GOSUB WrLCD
       char = %00000001           ' clear LCD
       GOSUB WrLCD
       HIGH RS                     ' LCD to character mode

       char = %01000000           ' address 0 of CG RAM
       GOSUB LCDcmd                ' prepare to write CG data
       FOR index1 = 0 TO 23        ' build 3 custom chars
          READ index1, char        ' get byte from EEPROM
          GOSUB WrLCD              ' put into LCD CG RAM
       NEXT index1
```

```
' -----[ Main Code ]------------------------------------------------

Start:  char = ClrLCD              ' clear the LCD
        GOSUB LCDcmd

        FOR index1 = 24 TO 39      ' write the character string
          READ index1,char         ' get char from EEPROM
          GOSUB WrLCD              ' write it
          PAUSE 50                 ' delay 50 ms--for fun only
        NEXT index1
        PAUSE 1000                 ' pause 1 second

        FOR index1 = 0 TO 15       ' cover 16 characters
          FOR index2 = 0 TO 4      ' 5 characters for animation
            char = %10000000 | index1' set DD RAM address
            GOSUB LCDcmd           ' move cursor to new addr
            LOOKUP index2,(0,1,2,1," "),char
            GOSUB WrLCD            ' write animation character
            PAUSE 75               ' delay between chars
          NEXT index2
        NEXT index1
        PAUSE 1000

        GOTO Start                 ' replay

' -----[ Subroutines ]----------------------------------------------

' Send command to the LCD
'
' Load char with command value, then call
'
'   Clear the LCD............ $01, %00000001
'   Home the cursor.......... $02, %00000010
'   Display control.......... (see below)
'   Entry mode............... (see below)
'   Cursor left.............. $10, %00010000
'   Cursor right............. $14, %00010100
'   Scroll display left...... $18, %00011000
'   Scroll display right..... $1C, %00011100
'   Set CG RAM address....... %01aaaaaa (Character Generator)
'   Set DD RAM address....... %1aaaaaaa (Display Data)
'
' Display control byte:
'
```

```
'    % 0 0 0 0 1 D C B
'                | | -- blink character under cursor (1=blink)
'                | |--- cursor on/off (1=on)
'                ------ display on/off (1=on)
'
' Entry mode byte:
'
'    %   0 0 0 0 0 1 X S
'                |   -- shift display (S=1), left (X=1), right (X=0)
'                ---- cursor move: right (X=1), left (X=0)
'
LCDcmd: LOW RS
        GOSUB WrLCD
        HIGH RS
        RETURN


' Write ASCII char to LCD
'
WrLCD:  Pins = Pins & %00010000        ' RS = 1, data bus clear
        outp = char / 16               ' get high nibble
        Pins = Pins | outp             ' output the nibble
        PULSOUT E, 1                   ' strobe the Enable line
        Pins = Pins & %00010000
        outp = char & %00001111        ' get low nibble
        Pins = Pins | outp
        PULSOUT E, 1
        RETURN
```

In the section Initialization all codes for the several pixel rows are saved in EEPROM. At the end of this section address 0 of CG ROM is set. That means that the first character will be defined by address 0. Next, reading EEPROM and writing CG RAM of the LCD controller follow. Address incrementation works automatically in this initialization. After writing the CG RAM the three user-defined characters are available with the ASCII codes 0, 1, and 2.

The program examples LCDDEMO1.BAS and LCDDEMO2.BAS were coded for a 16 × 1 LCD. If these programs control a 16 × 4 LCD, for example, then all display operations occur in the top row only. The following program example MYDEMO.BAS (listed partly) shows the changes required to control an LCD with a different size. We list only parts that differ from the last two program examples.

```
' -----[ Title ]-----------------------------------------------
'
' File...... MYDEMO.BAS
' Purpose... Stamp -> LCD (4-bit interface)
```

```
' Author.... Jon Williams (Adaptions to LM041 Claus Kühnel)
' Started... 16 July 1994
' Updated... 28 Sept 1994


' -----[ Program Description ]-------------------------------------
'
' This program demonstrates the various standard features of an LCD
' display that uses the Hitatchi HD44780 controller. The LCD used
' to test this program was the Hitachi LM041L (16x4).
'


...


' -----[ Revision History ]----------------------------------------
'
' 07-16-94 : Version 1.0 - compilation of code from last 3 months
' 09-20-94 : Version 2.0 - adaption to LM041L

' -----[ Constants ]-----------------------------------------------
'


...


SYMBOL  Z1 = $80          ' addr line #1 | 80H
SYMBOL  Z2 = $C0          ' addr line #2 | 80H
SYMBOL  Z3 = $90          ' addr line #3 | 80H
SYMBOL  Z4 = $D0          ' addr line #4 | 80H

' -----[ Variables ]-----------------------------------------------
'


...


' -----[ Initialization ]------------------------------------------
'
            EEPROM ("THE BASIC STAMP!")    ' preload EEPROM
            EEPROM ("A PIC16C56 knows")    ' preload EEPROM
            EEPROM ("P B A S I C from")    ' preload EEPROM
            EEPROM ("  Parallax Inc. ")    ' preload EEPROM

            Pins = %00000000               ' clear the pins
            Dirs = %00111111               ' set 0-5 as outputs
            PAUSE 20                       ' let the LCD settle
```

```
' Initialize the LCD (Hitatchi HD44780 controller)
'
LCDini: Pins = %00000011                    ' 8-bit mode
         PULSOUT E, 1
         PAUSE 5
         PULSOUT E, 1
         PULSOUT E, 1
         Pins = %00000010              ' 4-bit mode
         PULSOUT E, 1
         char = %00101000              ' set function for LM041
         GOSUB WrLCD
         char = %00001100              ' disp on, crsr off, blink off
         GOSUB WrLCD
         char = %00000110              ' inc crsr, no disp shift
         GOSUB WrLCD
         char = %00000001              ' clear LCD
         GOSUB WrLCD
         HIGH RS                       ' LCD to character mode


' -----[ Main Code ]-----------------------------------------------
'
Start:  FOR index = 0 TO 15
           read index, char           ' get char from EEPROM
            GOSUB WrLCD                ' write it
         NEXT index

         char = Z2                     ' address second line
         GOSUB LCDcmd
         FOR index = 16 TO 31
           read index, char            ' get char from EEPROM
            GOSUB WrLCD                 ' write it
         NEXT index

         char = Z3                      ' address third line
         GOSUB LCDcmd
         FOR index = 32 TO 47
           read index, char             ' get char from EEPROM
            GOSUB WrLCD                  ' write it
         NEXT index

         char = Z4                       ' address forth line
         GOSUB LCDcmd
         FOR index = 48 TO 63
           read index, char              ' get char from EEPROM
            GOSUB WrLCD                   ' write it
```

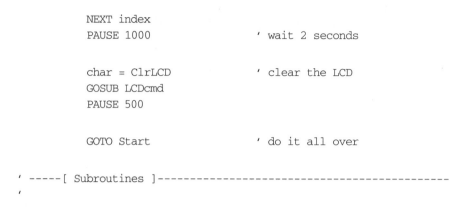

```
        NEXT index
        PAUSE 1000                  ' wait 2 seconds


        char = ClrLCD               ' clear the LCD
        GOSUB LCDcmd
        PAUSE 500


        GOTO Start                  ' do it all over


' -----[ Subroutines ]-------------------------------------------
'


. . .
```

[BS1]—Driving Seven-Segment LEDs Alpha-numerical displays and key-boards are important elements for interfaces between men and machine. While CRTs are the display units for the PC, for the microcontroller the important display devices are based on LCD or LED. Working with LCDs was described in the last chapter. Now we have a look at seven-segment LED displays, often used for numerical displays.

The specific application determines the appropriate number of seven-segment LED devices. For applications with the BASIC Stamp, display expenses should be minimized, due to the limited I/O capabilities.

The display driver device MAX7219 (MAXIM), our next application example, is a good compromise with reference to the expenses. The display driver MAX7219 can control up to eight seven-segment LED displays. In this example only two devices HDSP5523 (Hewlett Packard) with two displays each are controlled by the MAX7219. The interface to the BASIC Stamp is quite simple. A three-wire synchronous serial interface moves the data to be displayed from the BASIC Stamp into the internal shift register of the MAX7219. This shift register takes care of serial-to-parallel conversion and internal logic multiplexes in the display devices. Figure 7.36 shows the complete hardware with a BASIC Stamp I.

The serial data transmission from BASIC Stamp I to the display driver device MAX7219 happens via three wires:

CLK   Clock for synchronous serial data transmission

DIN   Input for serial data (digit and segments)

LOAD   Latch pulse for received data word (16-bit)

Via pin DIN, the display driver device receives a 16-bit data word containing the information here in Table 7.12:

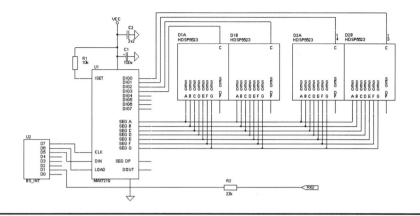

Figure 7.36
Driving the HDSP5523 with the display driver device MAX7219

Table 7.12
Contents of data word

Address	X	X	X	X	D11	D10	D9	D8
Data	D7	D6	D5	D4	D3	D2	D1	D0

The address is used for marking the display position for one of eight possible seven-segment LED displays by inputting a position between one and eight, but also for the addressing of control words for:

- Seven-segment decoding   Conversion of BCD into seven-segment code
- Intensity   Brightness control
- Scanlimit   Limitation of multiplexing to the number of displays used
- Shutdown   Power ON / OFF

For all these control words, the data field (D7 . . . D0) contains a certain argument. MAXIM's data sheet for the MAX7219 gives detailed information.

The display driver MAX7219 controls a maximum of eight characters of a seven-segment display. The MAX7219 is cascadable. In this case, all shift registers are cascaded to one big shift register for all display devices. The serial data stream must have the correct order so each shift register gets the correct data to the display.

For serial data transmission, the number of control lines is independent of the number of displays. The number of required wires is three.

In this application only four of the eight possible characters will be controlled by the use of a MAX7219. Figure 7.36 showed the connection of the

LED seven-segment displays HDSP5523. The low-active driver for digit 0 to digit 3 (Pin for common cathodes) works in multiplex with the Hi-active segment drivers for the segments A to G and the decimal point (DP). All decimal points are not connected.

The brightness of the display segments is programmed by the hardware. An internal current source is controlled by an external resistor of 10 kΩ connected with V_{CC} = 5 V. To avoid disturbances from pulses, V_{CC} is buffered with the capacitors C1 and C2.

The program TASTSER.BAS gets two characters via the serial interface and interprets them as a position; that segment switches on in this position. This input happens with a terminal program from the PC side to the BASIC Stamp. After each character you have to press the Enter key.

```
' -----[ Title ]------------------------------------------------
'
' File......tastser.BAS
' Purpose...display 4  digits with 2  seven-seg.displ. HDSP5523
' Author....Klaus Zahnert
' Started...
' Updated...


' -----[ Program Description ]-------------------------------
'       Two double-7-segment displays are connected with ser.-interf.
'       8 digit LED-display-driver. Synchr. serial transmission with
'       connections CLOCK,DATA,LOAD to stamp-ports.

' -----[ Revision History ]---------------------------------
'

' -----[ Constants ]-----------------------------------------
'
        symbol clk       = 7          'CLOCK-Connect MAX7219
        symbol load      = 6          'LOAD -Connect MAX7219
        symbol dout      = 5          'DATA -Connect MAX7219
        symbol RXD       = 1          'RS232 input from PC
        symbol baud      = n2400
        symbol shutdown  = $0C
        symbol decode    = $09
        symbol scanlimit = $0B

' -----[ Variables ]-----------------------------------------
'
        symbol word      = w5
        symbol data      = b10
```

```
       symbol adress    = b11
       symbol memo      = w0
       symbol si        = pin5
       symbol i         = b2
       symbol maxbit    = bit15


' -----[ Initialization ]------------------------------------------
'
       dirs  =  %11100000    '0 = input , 1 = output
       low   clk
       high  load

' -----[ Main Code ]-----------------------------------------------
'
start: adress = shutdown    'power up
       data = 1
       gosub wout
       adress = decode      'decode all digits
       data = $ff
       gosub wout
       adress = scanlimit   'display digit 0..3
       data = 3
       gosub wout
loop:  serin rxd,baud,#adress,#data    'adress 0...3 for digit 0...3
                                       'data  0 ... 9 display on
                                        digit

       gosub wout
       goto loop

' -----[ Subroutines ]---------------------------------------------
'
wout : memo = word
          'debug adress,data          'TEST

              for i = 1 to 16
                  si= maxbit          'actual bit to dataport for outp.
                  'debug si           'TEST
                  pulsout clk,10       'clock out
                  memo = 2 * memo 'shift left memo for actual bit
                                       'places in bit15
              next i
         pulsout load,10
         return
```

The subroutine wout contains the conversions routine for data saved in the variable word. A saved data word will be moved to the display starting with the

Figure 7.37
Timing serial interface

MSB. Figure 7.37 shows the timing of one serial transmission of one 16-bit data word to the MAX7219.

The serial data transfer starts with data bit D15 as MSB. Each rising edge on line CLK shifts in a data bit. If the data word is shifted in completely, it will be latched with the strobe on the line LOAD. Before the first data transfer, the initialization sets LOAD to Hi.

In our program example TASTSER.BAS, the transfer of all 16 bits works in a loop. In each circulation the MSB of the variable memo will be moved to the I/O line Pin5. Before a first output, the data word must be saved in this variable.

The line CLK is initialized with Lo. The command pulsout clk, 10 generates a short Hi pulse. With the rising edge of this clock pulse, the MAX7219 reads that data bit. In the first circulation the data bit D15 was clocked in. In the second circulation it should be D14. Multiplying by two shifts the contents of the variable memo one position to the left. Now we have in MSB the data that was placed on position D14 before multiplication.

In the subroutine wout two debug commands were included—marked here as comments. Activating these commands shows the program actions in this section. Because there is no handshake between the BASIC Stamp and the MAX7219, the program also runs without interface. In the debug window, all actions starting from the serial input from the PC to the serial output to the MAX7219 are demonstrated.

Before the program will circulate in the loop, there are three initialization steps. The conditions for shutdown, decoding mode and scanlimit (number of character positions) must be transferred.

This program with only four characters is good for many standard applications. Further enhancements are no difficult task.

[BS2]—Driving a 14-Segment LED Seven-segment displays are suitable for displaying numerical data. Even hexadecimal numbers can be displayed, although opinions differ as to the appearance of the hex numbers A to F. Increasing the number of segments helps to display better-resolved characters. Figure 7.38 shows a two-position 14-segment LED display, TSM7052 (Three-Five Systems, Inc.).

The TSM7052 features two 14-segment alphanumeric characters with on-board serial data input and parallel data-out LED drivers, designed to oper-

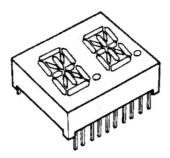

Figure 7.38
14-segment LED display

ate with minimal interface to the display's data source. The current through the LED is controlled by an external resistor.

The serial data transfer from BASIC Stamp to the display needs only three wires: for data input, data enable, and clock. The data format consists of a leading "1" followed by 35 data bits. All 35 data bits are latched after the 36th bit is complete, thus providing nonmultiplexed direct drive to the LED. Outputs change only when the serial data bits differ.

The position of the segments in the serial data stream is displayed in Table 7.13.

Table 7.13
Serial input sequence

Bit#	Digit#	Segment	Bit#	Digit#	Segment
1	2	A	18	1	D
2	2	B	19	1	E
3	2	C	20	1	F
4	2	D	21	1	G
5	2	E	22	1	H
6	2	F	23	1	K
7	2	G	24	1	M
8	2	H	25	1	N
9	2	K	26	1	R
10	2	M	27	1	S
11	2	N	28	1	T
12	2	R	29	1	DP
13	2	S	30	2	DP
14	2	T	31		Pin17
15	1	A	32		Pin1
16	1	B	33		Pin2
17	1	C	34		Pin3

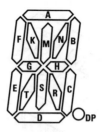

Figure 7.39
Segment designation

To allow two characters to display, all segments must be set or reset. Figure 7.39 shows the segment designation for this 14-segment LED display.

Table 7.14 shows the whole procedure for a day-of-the-week display. In the left column of Table 7.14 are the characters to be displayed with the 14-segment display for the day-of-the-week application. For Monday, you find the short form Mo (first and second row). All lightened segments must be set to "1." The resulting bit pattern gives a 14-bit code characterizing the defined alphanumeric character.

The following program example TSM7052.BS2 controls a TSM7052 LED display in an endless loop. One after the other the short forms of all days of the week are displayed.

Table 7.14
Defining different alphanumeric displays

Character	A	B	C	D	E	F	G	H	K	M	N	R	S	T	Code
M	0	1	1	0	1	1	0	0	1	0	1	0	0	0	1B28
o	0	0	1	1	1	0	1	1	0	0	0	0	0	0	0EC0
T	1	0	0	0	0	0	0	0	0	1	0	0	1	0	2012
u	0	0	1	1	1	0	0	0	0	0	0	0	0	0	0E00
W	0	1	1	0	1	1	0	0	0	0	0	1	0	1	1B05
E	1	0	0	1	1	1	1	0	0	0	0	0	0	0	2780
h	0	0	1	0	1	1	1	1	0	0	0	0	0	0	0DC0
F	1	0	0	0	1	1	1	1	0	0	0	0	0	0	23C0
r	0	0	0	0	0	0	0	1	0	0	0	0	1	0	0042
S	1	0	1	1	0	1	1	1	0	0	0	0	0	0	2DC0
A	1	1	1	0	1	1	1	1	0	0	0	0	0	0	3BC0

```
' -----[ Title ]-------------------------------------------------------

' File...... TSM7052.BS2
' Purpose... Example for display control of 14 Segment LED
' Author.... Claus Kuehnel
' Started... 13.10.95
' Updated...


' -----[ Program Description ]-----------------------------------------

' This demo shows display control for a 14 segment alphanumeric LED
' with two characters. Data transmission is synchron serial over a
' tree-wire interface.
' The required pattern for LED control is saveed in EEPROM. In this
' example the patterns of the days of the week are stored.
' For enhancements of the character set following this way are easy.


' -----[ Revision History ]--------------------------------------------

' 13.10.95: Version 1.0


' -----[ Constants ]---------------------------------------------------

null        CON 0
sda         CON 0
scl         CON 1
enable      CON 2


' -----[ Variables ]---------------------------------------------------

pattern     VAR    word
lft_chr     VAR    word
rgt_chr     VAR    word
temp        VAR    word
offset      VAR    byte
i           VAR    nib
col         VAR    nib


' -----[ Initialization ]----------------------------------------------

table DATA word $2dc0, word $0E00  ' pattern for "Su"
      DATA word $1b28, word $0ec0  ' pattern for "Mo"
      DATA word $2012, word $0e00  ' pattern for "Tu"
```

```
         DATA word $1b05, word $2780   ' pattern for "WE"
         DATA word $2012, word $0bc0   ' pattern for "Th"
         DATA word $23c0, word $0042   ' pattern for "Fr"
         DATA word $2dc0, word $3bc0   ' pattern for "SA"


' -----[ Main Code ]----------------------------------------------
'
start:     for i=0 to 6
             offset=i: col=0
             ' display left col in left character
             gosub read_pattern : lft_chr = pattern
             offset=i: col=1
             ' display right col in right character; mask for i=0..15
             gosub read_pattern : rgt_chr = pattern
             ' enable LED for data transmission
             low enable
             ' shift out start bit
             shiftout sda,scl,MSBFIRST,[$f\1]
             ' shift out right character
             shiftout sda,scl,MSBFIRST,[rgt_chr\14]
             ' shift out left charcter
             shiftout sda,scl,MSBFIRST,[lft_chr\14]
             ' shift out decimal points and bit31..bit34
             shiftout sda,scl,MSBFIRST,[null\7]
             ' disable LED
             high enable
             pause 500
           next
         goto start
end


' -----[ Subroutines ]----------------------------------------------
'
' read pattern from coding table in EEPROM
read_pattern:
         offset = (offset*4) + (col*2) + table
         read offset+1, temp
         pattern = temp*256
         read offset, temp
         pattern = pattern + temp
return
```

In the initialization section, the character table is built in the EEPROM. Each program line in this section marks one day of the week. Data in that table is accessible through an offset between 0 and 6.

After reading the pattern for the left and right characters, the serial data sequence must be built. The comfortable instruction shiftout of the new BASIC Stamp II allows quite a simple shift procedure.

The serial data sequence can be divided this way:

1. Startbit   ☞   1 Bit
2. Right character   ☞   14 Bit
3. Left character   ☞   14 Bit
4. Decimal points and digital outputs ☞   7 Bit

With four separate shift operations, the 36 bits of the whole serial data sequence can be transferred. After the backslash behind the variable name, the number of bits to transfer is given.

Because the decimal points and the digital outputs are not used, all bits after both characters are zero in this example.

The hardware for this application example is quite simple, too. Figure 7.40 shows the interface between BASIC Stamp II and the TSM7052 LED display.

7.2.6 Management of Time Information

[BS1]—Real Time Clock with Event Control   There are many ways to build a precise timer for a programmable real time clock. The internal clock of the BASIC Stamp or the frequency of a power line are usable. Much more accurate are time informations transmitted via long waves or satellites of the global positioning system.

Not so accurate, but accurate enough for many applications, is the quartz base used in most watches. Many quartz watches have hands driven by a stepper

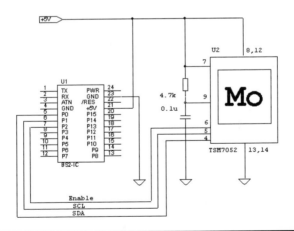

Figure 7.40
Controlling the TSM7052

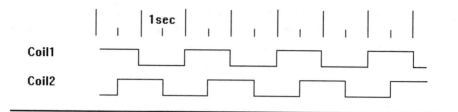

Figure 7.41
Timing on the stepper motor

motor. Small batteries with supply voltages between 1 V and 2 V feed the whole circuitry over a long time.

Wristwatches have stepper motors with one coil. Bigger clocks have stepper motors, usually with two coils. They are controlled with rectangle pulses with a phase of 90°. These pulses are controlled by the time clock.

Figure 7.41 shows the timing on the watch's stepper motor. This watch IC builds the timer for the BASIC Stamp. So a watch with faulty mechanics can serve as a timer. The expense for connecting the BASIC Stamp and the watch IC is very low. One transistor and two resistors amplify the coil voltage to about 1 V, and provide the signal range of an I/O pin. Figure 7.42 shows all of the hardware.

The supply voltage comes from the BASIC Stamp. Opening the watch and connecting two wires to a coil should be an easy task. The edge changes at the model used occur one time in a second. The resulting program for the BASIC Stamp is shown in the following listing (File UHR.BAS).

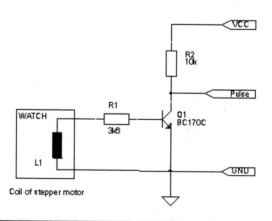

Figure 7.42
Generating timer clock

```
' -----[ Title ]-------------------------------------------------
'
' File......Uhr.bas
' Purpose...digital alarm-clock, controlled by stepping motor
'           pulse.
' Author....Klaus Zahnert
' Started...
' Updated...


' -----[ Program Description ]-------------------------------------
'      Stepping motor pulse of electr. watch is polled by stamp-
'      input for giving clock to time-register. Comparis. with
'      alarmsetting makes programmable alarm-output.
'


' -----[ Revision History ]---------------------------------------
'
'


' -----[ Constants ]-----------------------------------------------
'
        symbol al_line     = 7
        symbol ala_std     = 00        'preset alarm before RUN
        symbol ala_mnt     = 00        '=======================

' -----[ Variables ]-----------------------------------------------
'
        symbol clock = pin1
        symbol sec = b4
        symbol mnt = b5
        symbol std = b6
        symbol mark = bit0

' -----[ Initialization ]------------------------------------------
'
        dirs = %11110000
        low al_line 'alarm off
        mark = 00


        sec = 00 'preset time before RUN
        mnt = 00 '=====================
        std = 00

' -----[ Main Code ]-----------------------------------------------
'
```

```
start:   gosub anz
         gosub alarm

wait:    if clock = mark then wait 'loop waiting change
                                    'clock level
         mark = clock 'mark new level
         sec = sec + 1
         if sec = 60 then minute
         goto start

minute: sec = 0
         mnt = mnt + 1
         if mnt = 60 then hour
         goto start

hour:    mnt = 0
         std = std + 1
         if std = 24 then day
         goto start

day:     std = 0
         goto start

' -----[ Subroutines ]-------------------------------------------------
'

anz:     debug cls,#std,": ",#mnt,": ",#sec 'any user display
         return

alarm:   if std = ala_std and mnt = ala_mnt then alout
                                              'Test Alarm
         return

alout:   pulsout al_line,1000 'Alarm 10ms
         return
```

The entry point start will be reached from different program points in an endless loop. The first instructions after start execute some typical user subroutines for display and alarm detection and execution.

In a one-line-loop, the I/O pin1 (clock) will be asked for its level. The variable mark is referenced, storing the level from the last run through the loop. On the start mark will be reset.

If wait and mark are antivalent then an edge change has occurred and the seconds must be incremented. After incrementation of the seconds a test will be run to see if other incrementations (minute, hour, day) are required yet.

With the subroutine `anz` the user can implement his or her own display routine. The display could be an LCD, LED, or the PC's screen, via `debug` instruction. In the listing above (File `UHR.BAS`) the changes can be inspected.

In the subroutine `alarm` the actual values for minute and hour are tested against programmable time data for an alarm. In the case of equivalence, the alarm routine `alout` will be executed. In this example, a 10 ms pulse to an alarm output could activate a relay.

The program is ready for your modification. Due to the external generation of the second, we have reserves in internal RAM for additional program memory. Further program routines can be implemented. Please note that the longest execution time may not exceed the period of one second. If this happens the time error grows in increments of precisely one second.

To simplify matters, dialogs for input of a start and an alarm time do not exist. Both times must be indicated in the source immediately before downloading this program. If the time cells are loaded with zero, this timer works as a relative timer.

[BS2]—Real Time Clock DS1202   In recent years, the connections between microcontrollers and peripherals have improved synchronous transmission for several components. For example:

- Analog-digital and digital-analog conversion.
- Memory.
- Port expanding.
- Sensors for temperature, pressure, humity, and other more.
- Time keeping.

Only two or three wires are necessary to connect components. They are:

- Data lines in one or both directions.
- A clock to mark the change of data bits.
- A control line for reset or transmission enable (only in three-wire connections).

It goes without saying that a ground connection is necessary.

Protocols such as I²C, SPI, and Microwire are important ones, based on firm standards, and are widely used. And BASIC Stamp I can support different synchronous protocols, but you must write a package of statements for synchronous serial in and out.

For BASIC Stamp II, Parallax implemented the instructions `shiftin` and `shiftout`. See Chapter 5, the section called PBASIC Instruction Set, for definitions. Behind both instructions are assembler routines.

So we have the first advantage: There are not many statements in PBASIC, so we save capacity in EEPROM. Second, we find that these instructions work more quickly. (In our first example we have roughly calculated a factor of 10.)

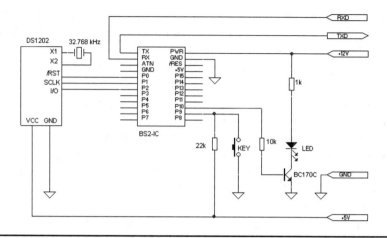

Figure 7.43
Connect DS1202 to BS2

Now we start with the timekeeper DS1202. With a connected 32,768 kHz crystal, typical in clocks, there is a complete clock and calendar. Optional are 24 static RAM cells with 8 bits. Their autonomous use could be a welcome RAM expansion to the BASIC Stamp registers. But keep in mind: Each cell holds one byte, not a word!

The helpful features and their use that are described in this chapter extend only to two examples. For further information, see the DALLAS documentation, so you can expand your own work.

Figure 7.43 shows the connections between BASIC Stamp II and Dallas DS1202. There are only three wires for bi-directional data transmission: I/O for data, SCLK for stamp-generated clock, and a low-active reset by /RST. Optionally, we may use our two serviceable tools: LED on Pin15 and key on Pin14. We think it is a help for debugging and for run control. We have also used it for measuring with an oscilloscope on Pin14 to show run-time.

Read and Write a RAM Cell A first example for shiftin and shiftout is loading RAM0 with different values of a byte. Then we read it and show the read value with debug in the PC's debug window.

Before we used the new instructions shiftin and shiftout we used subroutines—a must for BASIC Stamp I. We were happy to see the same effect. The loaded byte was read by debug in the debug window. This "self-programmed shiftin and shiftout" is the transfer of the pulse diagram of DS1202's serial protocol, shown in Figure 7.44, to single program steps in the subroutines shinbyte and shoutbyte in the following listing (File RAM0TST1.BS2).

/RST = Hi enables the byte transfer, with LSB always first.

BASIC Stamp

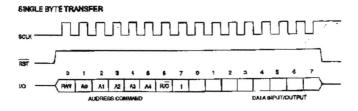

SINGLE BYTE TRANSFER

Figure 7.44
Single byte transfer for DS1202

For WRITE in DS1202, the LSB is output from BASIC Stamp before the first clock pulses out. Following clocks, write in the actual next bits, setting to data input of DS1202 by BASIC Stamp II before output.

For READ from DS1202, the LSB appears after /RST goes Hi. Following clocks switched to next bits, generated by DS1202 on I/O-line.

Byte transmission stops after transmission of two bytes with /RST going Lo. We don't care for the burst mode transfer of DS1202; it is possible to transmit some bytes together, like a package, but to demonstrate shiftin and shiftout, it is not the way. If necessary, take multiple instructions of shiftin and shiftout. The instructions shiftin and shiftout are only allowed with bit counts from 1 to 16; 8 bit is the default.

Every transmission from or to DS1202 starts with the output of one byte, addressing and controlling the following data byte. To address and read or write RAM23, the address byte is as shown here, in Table 7.15.

```
' -----[ RAM-test  DS1202 ]-----------------------------------------
' File...... ram0tst1
' Purpose... use of shiftin/out - statements for read and
'            write IC DALLAS DS1202
' Author.... Klaus Zahnert
' Startet... 12.07.95
' updatet...
```

Table 7.15
Access to RAM cell

Bit 7	6	5	4	3	2	1	Bit 0
1	1	1	0	1	1	1	Rd/WR
Control	RAM			Address			

```
' -----[ Program Description ]--------------------------------------
'
' Byte-in RAM0  ==>Byte-out RAM0


' -----[ Constants ]------------------------------------------------
neg_rst       con 0        'hardware- connections
sclk          con 1        '"only 3 wires + GND"
io            con 2


ctrwritepr    con $8E      'writeprot. control byte
outnwritepr   con $00      'writeprot. cancel f. RAM
ctrwriteram0  con $C0      'RAM0 ctrl write
outexample    con 132      'Value set for example  CHANGE !!!
ctrreadram0   con $C1      'RAM0 ctrl read


' -----[ Variables ]------------------------------------------------
mem           var  byte    'temporary cell
bit_nr        var  byte    'index for loop
reg_adr       var  byte    'first byte to DS..
outbyte       var  byte    'second b. to DS.. f. wr.DS..


' -----[ Initialisation ]-------------------------------------------
dirh = %10000000           'LED-Anz. an p15,key an p14
dirl = %00000011           '/rst,sclk ==> Ausgänge
low  neg_rst               'preset signal
low  sclk                  'preset signal


' -----[ Main Code ]------------------------------------------------
'_____ read-in to  RAM/DATA0 of  DS1202_____

start:    reg_adr = ctrwritepr      'RAMx write enable
          outbyte = outnwritepr
          gosub w_out

          reg_adr = ctrwriteram0    'out example-byte to DS
          outbyte = outexample
          gosub w_out

debug "end in",cr                   'optional message
gosub brk
'_____ read-out from  RAM DATA0 DS1202 _____

          reg_adr = ctrreadram0     'in example-byte from DS..
          gosub w_in
```

```
'===================================================================
debug cr, "RAM0-data = ",dec mem,cr  'display RAM0
'===================================================================
end


' -----[ Subroutines ]-----------------------------------------

lamp: high 15                    'LED-driver on pin 15
      pause 1000                 '... only for test
      low 15
      pause 1000
      return


brk:  if in14 = 1 then brk       'breakpoint,go on with
                                 'key on pin 14...only f.test
      debug cr, "follow the program",cr
      pause 500
      return
'_____

' programmed "SHIFTOUT" /"SHIFTIN" for one byte with the
' shown pin-connections  and transfer with "mem"-cell

shoutbyte:   output io               'direction out2
             for bit_nr = 0 to 7
                 out2 = mem.bit0      'io - Data - pin, lsb first
                 pulsout sclk,100     ' clock-pulse
                 mem = mem >>1        'mem shiftright  (div 2)
             next
             return


shinbyte:    input io                 'direction in2
             mem = 0                  'clear mem
             for bit_nr = 0 to 7
               mem = mem >>1           'mem shiftright (div 2)
               mem.bit7 = in2          'input mem.bit msb
               pulsout sclk,100        'get next databit  from
                                       'DS1202 to pin 2
             next
             return
'_____

W_out: mem = reg_adr
       high neg_rst
           'gosub shoutbyte                          'CHANGE
           shiftout io,sclk,lsbfirst,[mem]           'IT !!!
```

```
              mem = outbyte
              'gosub shoutbyte                        'CHANGE
              shiftout io,sclk,lsbfirst,[mem]          'IT !!!

         low neg_rst
         return

W_in:    mem = reg_adr
         high neg_rst
              'gosub shoutbyte                        'CHANGE
              shiftout io,sclk,lsbfirst,[mem]          'IT !!!

              'gosub shinbyte                         'CHANGE
              shiftin io,sclk,lsbpre,[mem]             'IT !!!

         low neg_rst
         return
'_____
```

The subroutines W_in and W_out near the end of the program listing include the two alternative statements for serial transmission. Pick one of them with "comment out." If you have an oscilloscope or a counter you can measure the run-time for each choice. Clamp the actual statement with LED ON/OFF, for example:

```
high 15
gosub shoutbyte
low 15
```

The pulse on LED measured by oscilloscope is approximately 20 ms for shoutbyte and 2 ms for shiftout, with a ratio of 10 : 1. For this measurement we used the shortest programmable pulsout with pulsout sclk, 1 in the subroutine shoutbyte.

Alarm Clock   Seven byte-registers are used in DS1202 for timekeeping and calendar. They can be used like free programmable registers, described before. The difference is that they are in a chain with overflow to one another, clocked by crystal for giving the second, minute, hour, date, month, day, and year.

You can preset all these registers with actual values after sending a no-write-protect command. The clock will start with these pre-set values after the write-protect command. The next listing (File CLOCK3.BS2) shows the program. We only need the cells for second, minute, and hour. You can expand for whole calendar and clock by yourself.

```
' -----[ alarm-clock with  DS1202 ]-------------------------------
'File...... clock3.bs2
'Purpose... use of shiftin/out - statements for clock in
'           IC DALLAS DS1202
'Author.... Klaus Zahnert
'Startet... 12.07.95
'updatet...


' -----[ Program Description ]------------------------------------
'RAM - cells for writeprotect sec, min, hour  are preset
'with actual values. Read-out of actual time from this
'cells by debug. Optional alarm-message to pc-speaker,
'started by successfull comparisation between presetted
'alarm-time with actual time.


' -----[ Constants ]----------------------------------------------
neg_rst        con  0       'hardware- connections
sclk           con  1       '" only 3 wires + GND"
io             con  2

ctrwritepr     con  $8E     'writeprot. control byte
outwritepr     con  $80     'writeprot. enable for RAM
outnwritepr    con  $00     'writeprot. cancel for RAM

ctrwritesec    con  $80
outsec         con  $00     'second = 00 const.

ctrwritemint   con  $82
outmint        con  $57     'minute 00 ... 59  act. PRESET !!!

ctrwritehour   con  $84
outhour        con  $11     'hour   00 ... 23  act. PRESET !!!

ctrreadsec     con  $81     'read contr. byte  sec
ctrreadmint    con  $83     '                  min
ctrreadhour    con  $85     '                  hour

alarmmint      con  $00     'minute 00 ... 59  alarm PRESET !!!
alarmhour      con  $12     'hour   00 ... 23  alarm PRESET !!!

'-----[ Variables ]----------------------------------------------
mem            var  byte        'temporary cell
bit_nr         var  byte        'index for loop
reg_adr        var  byte        'first byte to DS..
outbyte        var  byte        'second b. to DS.. f. wr.DS..
```

```
ss              var  byte        'seconds display
ssnew           var  byte        'actual poll seconds
mm              var  byte        'minutes display
hh              var  byte        'hours display


'-----[ Initialisation ]---------------------------------------
dirh = %10000000                 'LED-Anz. an p15,key an p14
dirl = %00000011                 '/rst,sclk ==> outputs
low  neg_rst                     'preset signal
low  sclk                        'preset signal


'-----[ Main Code  ]-------------------------------------------

start: If IN14 = 1  then readout  'if no keypressed, clock
                                  'ignore preset-time-values


'_____ read-in time of DS1202_____

readin: reg_adr = ctrwritepr     'RAMx write enable
        outbyte = outnwritepr
        gosub w_out

        reg_adr = ctrwritesec
        outbyte = outsec         'set second
        gosub w_out

        reg_adr = ctrwritemint
        outbyte = outmint        'set minute
        gosub w_out

        reg_adr = ctrwritehour
        outbyte = outhour        'set hour
        gosub w_out

        reg_adr = ctrwritepr     'RAMX write protect
        outbyte = outwritepr
        gosub w_out

debug "end preset time in",cr    'optional message
pause 1000

'_____ read-out time  DS1202 _____
```

```
readout:    reg_adr = ctrreadsec                 'read second
            gosub w_in
            ssnew = mem
            If ssnew = ss then readout
            ss = mem

            reg_adr = ctrreadmint                'read minute
            gosub w_in
            mm = mem

            reg_adr = ctrreadhour                'read hour
            gosub w_in
            hh = mem
'================================================================
debug cr, "TIME = ",HEX2 hh,":",HEX2 mm,":",HEX ss,cr
'================================================================
gosub lamp                        'visual watcher with LED
gosub alarm                       'alarm if condition true
goto start                        'loop
end


'-----[ Subroutines ]-----------------------------------------


lamp:    high 15                       'LED-driver on pin 15
         pause 10
         low 15
         return
'
         _____


         W_out:  mem = reg_adr              'input: reg_adr,outbyte
         high neg_rst
             shiftout io,sclk,lsbfirst,[mem]
             mem = outbyte
             shiftout io,sclk,lsbfirst,[mem]
         low neg_rst
         return
W_in:    mem = reg_adr                      'input: reg_adr
         high neg_rst
             shiftout io,sclk,lsbfirst,[mem]
             shiftin io,sclk,lsbpre,[mem]
         low neg_rst
         return                             'result: mem
'
         _____
```

```
alarm:    if mm=alarmmint and hh=alarmhour then al_on
          return
al_on:    debug cr,"A L A R M ", 07, cr          'display alarm
          return                                  'with PC-speaker
'_____
```

The minutes and hours of the actual time are preset as constants in the program. There is no programmed input for this example. Also, you can preset an alarm time. This feature is a little comfort for this clock. In the same way you can initialize any other events.

The clock runs after starting the program and preset values. In order to run with preset values after the first program start, you skip preset in run statements. If you choose the switch in Figure 7.43, preset values are going to register.

A little "heartbeat" is made with an LED pulse of 10 ms in each second.

7.2.7 I/O Extension with I²C Circuits

The limited I/O resources of the BASIC Stamps can be enhanced with I²C-bus devices. These devices have a serial bi-directional two-wire bus for communication, and they place different peripheral functions at the user's disposal. The devices PCF8591 and PCF8574/8574A will be considered. The PCF8591 is an 8-bit AD/DA converter and the PCF8574/8574A is an 8-bit I/O expander.

AD/DA-Converter PCF8591   The PCF8591 contains an analog-to-digital converter which workes by successive approximation. Furthermore, it has four input channels and one digital-to-analog converter on the same chip. Both converters work with a resolution of eight bits. The supply voltage must be between 2.5 V and 6 V DC. With the input multiplexer, numerous configurations of the four input channels are possible. A track&hold device is on-chip, too.

A reference voltage must come from an external source. If the supply voltage is stable enough, it can also serve as a reference voltage.

A clock is needed for the analog-to-digital conversion and the adjustments of the buffer amplifier. Whether an external or an internal oscillator generates the required clock depends on the logic level at pin EXT.

Figure 7.45 shows the pins of the analog-to-digital converter PCF8591. In the upper part of the picture, all pins of the AD/DA converter, including pins for reference and supply voltages and pins for the oscillator, can be found. Three address lines lead to the pins A2 to A0. Up to eight identical devices can operate on the same I²C bus. The remaining pins build the I²C bus interface.

I/O-Expander PCF8574/8574A   The PCF8574/8574A is an I/O expander with eight quasi-bi-directional I/O lines P0 to P7. The outputs are designed as

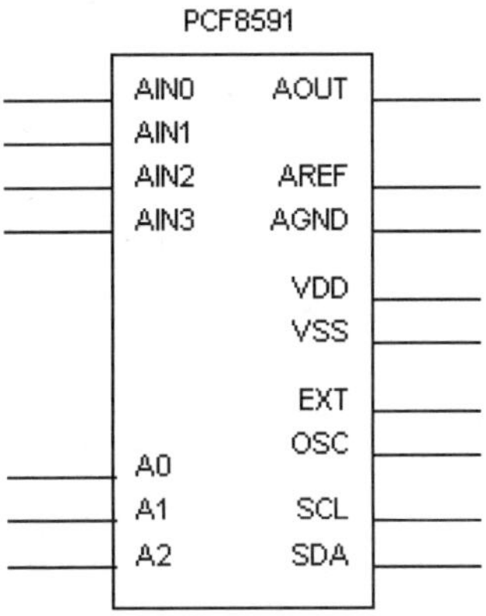

PCF8591

```
              PCF8591
        ┌──────────────────────┐
────────│ AIN0          AOUT    │────────
────────│ AIN1                  │
────────│ AIN2          AREF    │────────
────────│ AIN3          AGND    │────────
        │                       │
        │               VDD     │────────
        │               VSS     │────────
        │                       │
        │               EXT     │────────
        │               OSC     │────────
        │ A0                    │
────────│ A1            SCL     │────────
────────│ A2            SDA     │────────
────────│                      │
        └──────────────────────┘
```

Figure 7.45
PCF8591 pins

open-drain outputs and can directly drive LEDs with a current of 25 mA maximum. The interrupt output /INT signals that there are changes on the I/O port. The supply voltage is again between 2.5 V and 6 V DC. A2 to A0 are three address pins again, too. Eight PCF8574s can operate on the same I²C bus. For a further enhancement, the PCF8574A provides eight additional expander devices. The remaining pins are again the I²C bus interface. Figure 7.46 shows the pinning of both I/O expanders.

I²C-Bus The I²C bus is a bi-directional two-wire bus for communication between different modules. The two wires are the clock SCL and the data line SDA. Via the

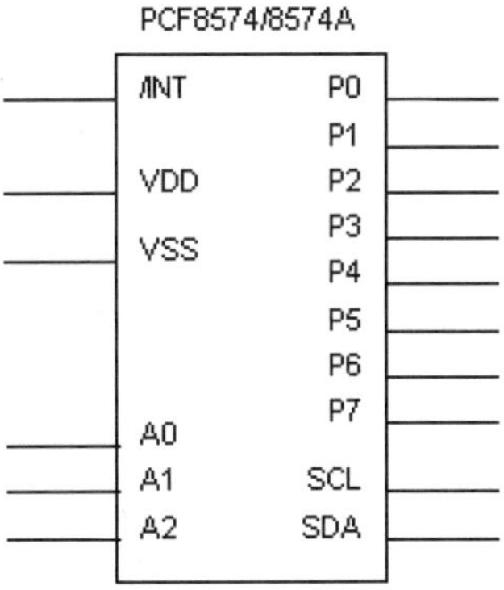

```
            PCF8574/8574A
        ┌──────────────────────┐
────────│ /INT          P0     │────────
        │               P1     │────────
────────│ VDD           P2     │────────
        │               P3     │────────
────────│ VSS           P4     │────────
        │               P5     │────────
        │               P6     │────────
        │               P7     │────────
────────│ A0                   │
────────│ A1            SCL     │────────
────────│ A2            SDA     │────────
        └──────────────────────┘
```

Figure 7.46
PCF8574/8574A pins

clock line SCL, each device connected to the bus gets the clock for reading and writing of the serial data. The data line SDA is bi-directional and handles the data exchange. The software protocol always defines the direction of data exchange.

For a data exchange with an I²C-bus device, the master sends an address byte first, and the communication with the addressed device (slave) starts. The structure of the address bytes for the devices considered here is shown in Table 7.16.

With the address bits A2 to A0, it is possible to control the access for more than one device. In general the device address will be defined through DIP switches, jumpers, or soldered bridges near the device. The data exchange with the addressed device begins after the address byte has been sent. Figure 7.47 shows the software protocol for possible accesses to the devices.

The sections marked with START and STOP are characterized by a certain phase layer of the signals SCL and SDA. The individual bytes will be separated by an acknowledge ACK sent from the addressed device to the master for error-free data exchange. Figure 7.48 shows the different signal patterns on the I²C bus.

There is a peculiarity in the digital-to-analog conversion: In Figure 7.47, a control byte follows the address byte. This control byte serves as the control of all device functions. Table 7.17 describes the bits of this control byte.

The bits in the control byte need explanation. The analog output enable flag AOE switches the output of the digital-to-analog converter in an active (AEO = 1) or in a high impedance (AOE = 0) state. The bits CONF1 and CONF0 make possible four configurations of the input multiplexer. If the auto increment flag is set (AI = 1), then the input channel will be incremented automatically after any conversion. For that reason, cyclic queries of some input channels are easily programmable. The address bits AN1 and AN0 select the input channel for conversion.

As Figure 7.46 shows, additional data bytes can follow a first data byte. Address and control bytes are not required for each digital-to-analog conver-

Table 7.16
Addresses of PCF8591/74/74A

Device	D7	D6	D5	D4	D3	D2	D1	D0
PCF8591	1	0	0	1	A2	A1	A0	R = 1; W = 0
PCF8574	0	1	0	0	A2	A1	A0	R = 1; W = 0
PCF8574A	0	1	1	1	A2	A1	A0	R = 1; W = 0

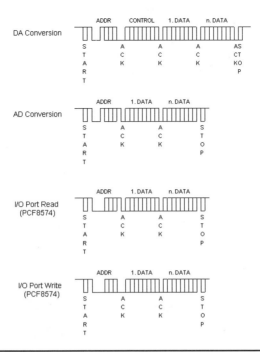

Figure 7.47
Software protocol for accessing the devices PCF8591 and PCF8574/8574A

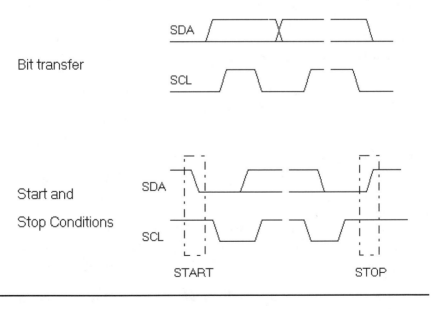

Figure 7.48
Signal patterns on I²C bus

Table 7.17
Contents of control byte

Control byte							
D7	D6	D5	D4	D3	D2	D1	D0
0	AOE	CONF1	CONF0	0	AI	AN1	AN0

sion. One or more digital-to-analog conversions end with the STOP condition of the I²C protocol.

The analog-to-digital conversion is very similar. The address byte characterizes the set bit D0 that read accesses to the PCF8591 follow. After that, analog-to-digital conversions will always start with those parameters written to the control register. The conversion will start with the transmission of the first bit. The conversion's result will be sent to the master with the next read operation. Reading the n-th data byte begins the (n + 1)-th analog-to-digital conversion. One or more analog-to-digital conversions end with the STOP condition of the I²C protocol, too.

Reading and writing the I/O expander happens in the same way. No control byte is needed; only address and data bytes exist. Bit D0 characterizes the read or write operation. Reading the expander means reading the I/O port and sending the data from PCF8574 to the master. For writing, the data direction changes.

Now that the BASIC Stamp enhancement has been discussed, the complete circuit design should be considered.

[BS1]—I/O-Module Figure 7.49 shows a BASIC Stamp enhanced to an I/O module. To simplify matters the whole module will be fed by a stable external supply voltage of 5 V DC. If this voltage is stable enough, then it can also serve as a reference voltage.

Pin EXT of the PCF8591 device is grounded. This means that the internal oscillator works and no other components are needed for clock generation. The oscillator frequency, in a range between 750 kHz and 1.25 MHz, can be found at pin OSC.

The I/O expander has no peculiarities. The interrupt output of the PCF8574 device leads to the BASIC Stamp's Pin2. If this input is polled, then a changed port at the PCF8674 device will be established and a query can be started.

[BS1]—Sample Applications with the I/O Module Three software examples will show the function of the AD/DA converter PCF8591 and of the I/O

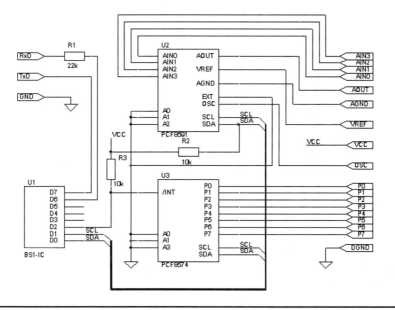

Figure 7.49
I/O Module with BASIC Stamp I

expander PCF8574. All software examples work with the hardware shown in Figure 7.50. That is why the definition and initialization parts of all three programs are basically equal.

The four potentiometers R9 to R12 lead to the four analog inputs of the PCF8591 device. The voltage from the potentiometers is therefore conversable. The analog output serves as an unbuffered output for oscilloscopes and/or multimeters.

The whole circuitry requires an external stabilized voltage supply of 5 V DC. The reference voltage $V_{REF} \approx 2.6$ V is generated by a resistor divider from the supply voltage V_{CC}.

For inspecting the digital I/O, two keys and two LEDs give some help. The limited resources of the BASIC Stamp force consideration of the available memory. In addition, such matters as the program structure and keeping the time conditions must be considered.

[BS1]—Sine Output. The BASIC Stamp I has no implemented sine function, so function values for a digital-to-analog converter must be calculated or taken from a table. Here, all function values of one period were saved in a table. Saving the values between 0 and 90° would be a further possibility. With the same need of memory, a higher resolution would be achieved, but the program would be more complex.

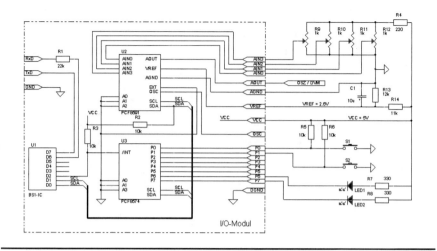

Figure 7.50
Peripheral components of the I/O module

The next listing (File SINOUT.BAS) shows the generation of a sine wave
and the required communication between BASIC Stamp and PCF8591.

```
' -----[ Title ]----------------------------------------------------

'
' File......  SINOUT.BAS
' Purpose...  Output of continuous sinuswave by PCF8591
' Author....  Klaus Zahnert
' Started...  6.10.94
' Updated...


' -----[ Program Description ]--------------------------------------
'
' Data of one period of a sinus wave are stored in EEPROM and
' converted into voltage signal by PCF8591.


' -----[ Revision History ]-----------------------------------------
'
'
'


' -----[ Constants ]------------------------------------------------
'
symbol scl  = 1          'clock              -output
symbol sda  = 0          'data               -output
```

```
symbol rxd  = 6          'rs232 from PC       -input
symbol txd  = 7          'rs232 to   PC       -output
symbol int  = 2          '/INT from PCF8591  -input ,not used
symbol baud = N2400


' -----[ Variables ]---------------------------------------------
'
symbol iobyte  = b0    'working reg. ser.<===>par.
symbol adress  = b2
symbol control = b5
symbol i       = b6    'common variable, special for loops
symbol j       = b7
symbol stack1  = b8
symbol stack2  = b9


' -----[ Initialization ]----------------------------------------
'
'table of one period sinus with increment 15 for 0 to 360

'Sinus v. Winkel    0   15   30   45   60   75   90 105 120 135 150 165
    eeprom 0,       (128,161,192,218,239,252,255,252,239,218,192,161)

'Sinus v. Winkel   180 195 210 225 240 255 270 285 300 315 330 345
    eeprom          (128, 95, 64, 38, 17,  8,  0,  8, 17, 38, 64, 95)


' -----[ Main Code ]---------------------------------------------
'
'      BSAVE              only while syntax-test without STAMP

       dirs  = %10000011            ' preset port direction for
                                    ' scl,sda,rxd,txd,int
       high    scl                  ' preset scl,sda for inactivity
       high    sda
       adress = %10010000           ' PCF8591 write for subadress %000
       control= %01000100           ' 4-chan.switched ADU,1*DAU

sinus: gosub start          '
       iobyte = adress      ' not interrupted chain
       gosub trans          ' of followed data
       iobyte = control     '
       gosub trans

       for i=0 to 23        ' one period
             read i,iobyte
```

```
                              ' debug i,iobyte
                              ' ==============
              gosub trans
              next i
        goto sinus             ' endless output


' -----[ Subroutines ]----------------------------------------
'
start:  low sda                'only once, beginning data transfer
        low scl
        return

stop:   high scl               'not used for continous transmission
        high sda
        return

trans:  for j = 0 to 7     ' with actual databyte filled in iobyte
            pin0 = bit7    ' copy actual bit from iobyte: MSB to SDA
            pulsout scl,1 ' highact. clock while act. bit on SDA
            iobyte = 2 * iobyte ' next bit with shift right in iobyte
                           ' debug j,pin0
                           ' pause 200
                           ' ===========
        next j

        dirs = %10000010  ' sda going input for recieve acknowl.
        high    scl
        if pin0 = 1  then messg   ' test aknowledge
        low     scl
        dirs = %10000011   ' sda returns to output
        low sda
        return

messg:  serout txd,baud,(" ACKN = 1")
                           debug pin0
                           '=========
        return
```

The declarations of the I/O ports are made according to the hardware in Figure 7.50. Constants and variables are separately declared. The possibility for bit addressing of word W0 allows a very simple serial data handling. The required bit stream will be generated by alternating bit shift and bit I/O.

Sample points of the sine function are saved in EEPROM. With 24 sample points, the phase step is 15°.

The first command in the main program marked with MAIN is BSAVE. The leading ' is only a comment, with no effect. Without this the leading character, BSAVE is a compiler directive. As a result of the compilation, tokens are normally generated for downloading in the BASIC Stamp's EEPROM. With BSAVE these tokens are saved instead in a file CODE.OBJ, an image of the data thought for download. We will find this file CODE.OBJ in the same subdirectory as our editor/compiler STAMP.EXE, and can analyze the memory needs of this program. In the tool chapter we described the program STMPSIZE to get a memory dump of an application program.

Figure 7.51 shows the memory dump of the program SINOUT.BAS generated with the program STMPSIZE.

In the lower area of addresses (top end), the sine table is saved. This area is called the user area. The program grows from higher addresses (bottom end) to this area. With this method, it is easy to get a quick impression of the available resources of the BASIC Stamp I. Remember, for BASIC Stamp II this tool is part of the development system.

Next, the start conditions for the I²C bus will be set. The endless loop marked with the label sinus outputs one period of the sine function now. The listing of the program SINOUT.BAS shows the details of the software protocol.

After the start message is sent, the address of the PCF8591 device and information about further write operations will be sent. The control byte (ctrl = %01000100) following the address byte enables the output of the digital-to-analog converter and configures the analog-to-digital converter. The details of

Figure 7.51
Memorydump

configuration are not important here. Now 24 data bytes can be sent to the digital-to-analog converter before this cycle is closed with the stop condition. This procedure will repeat endlessly.

The subroutine `trans` is responsible for sending a byte to the I²C bus. The variable `iobyte` delivers the byte to the subroutine. The serial transmission starts with the MSB (= Most Significant Bit) (pin0 = bit7). After the clock pulse (`pulsout scl,1`) is output with a duration of 10 μs, the rest of the data byte will be shifted one character to the left (`iobyte = 2*iobyte`) so that this procedure can be repeated.

After each byte is transferred, an acknowledge bit will be sent from the receiver to the transmitter, independent of which is master and which is slave. At this moment a sending master has to change its SDA line to an input. The received acknowledgement from any PCF device generates a message to the host for ACK = 1. The subroutine `message` does the job.

The acknowledge bit provides the typical multi-master operation of the I²C bus with changed sources for the start-stop handling of the embedded byte transmission. In the next examples, the BASIC Stamp is always the master. In addition to the evaluation of the acknowledge bit, pay careful attention to the bus timing between SDA and SCL in the ninth clock of the byte transmission.

In a multi-master application it is absolutely necessary to check if the bus is inactive (SCL = SDA = Hi) before a start condition is set. Because we work with only one master, such checks are not required.

[BS1]—Analog Input The opposite of the described output function is reading the analog-to-digital converter from the PC8591 device. The next listing (File ANAIN.BAS) shows a program that reads the analog-to-digital converter and transmits the results via RS232 to a connected host.

```
' -----[ Title ]-------------------------------------------------
'
' File..... ANAIN.BAS
' Purpose.. Analog input with I2C - interconnection.
' Author....Klaus Zahnert
' Started...04/94
' Updated...


' -----[ Program Description ]----------------------------------
'
' Using PCF8591 with I2C - Conn. to four-chan.autoincremented ADU.
' Periodically scanned values are transmitted to PC with RS232-conn.
'
' -----[ Revision History ]-------------------------------------
'
'
```

```
' -----[ Constants ]-----------------------------------------

        symbol sda = 0          'data                -output,input
        symbol scl = 1          'clock               -output
        symbol int = 2          '/INT from PCF8574    -input
        symbol rxd = 6          'RS232 from PC        -input
        symbol txd = 7          'RS232 to    PC       -output

        symbol baud         = N2400      'serial transmission rate
        symbol PCF8591_rd   = %10010001 'PCF8591 read  for subadr %000
        symbol PCF8591_wrt  = %10010000 'PCF8591 write for subadr %000
        symbol PCF8591_ctrl = %01000100 '4-chan.switched ADU,1*DAU
        symbol sda_in       = %10000010 'STAMP dirs preset
        symbol sda_out      = %10000011

' -----[ Variables ]-----------------------------------------

        symbol iobyte = b0    'working register ser. <====> par.
        symbol i      = b1    'i,j = common var.,special f. loops
        symbol j      = b2

' -----[ Initialization ]------------------------------------
        high scl              'preset scl,sda for inact. interconn. I2C
        high sda

' -----[ Main Code ]-----------------------------------------

'BSAVE                         'only while syntax test without STAMP
        gosub start           'Init PCF8591 for four single channel
                              'autoincrem. ADU and one chan. DAU
                              'with const. value
        iobyte = PCF8591_wrt
        gosub trout
        iobyte = PCF8591_ctrl
        gosub trout
        iobyte = 25           'DAU constant value
        gosub trout
        gosub stop

loop:   gosub start           'four-channel single-mode DAU with
        iobyte = PCF8591_rd   'serial transmission to PC
        gosub trout
        gosub trin
        pause 500
        debug iobyte
```

```
      serout TxD,baud,(iobyte)
      gosub stop
      goto loop                'endless loop


' -----[ Subroutines ]----------------------------------------
'
start: low sda                 'begin transmission
       low scl
       return


stop:  high scl                'terminates transmission
       high sda
       return


trout: dirs = sda_out
       for j = 0 to 7          'with actual databyte filled in iobyte
             pin0  = bit7       'copy actual bit from iobyte: MSB to SDA
             pulsout scl,10     'highact. clock while act. bit on SDA
             iobyte = 2*iobyte  'next bit with shift left in iobyte
                                'debug j,pin0
                                'pause 200
                                '===========


       next j
       gosub ackn
       return


trin:  dirs = sda_in
       for j = 0 to 7
             iobyte = 2*iobyte
             high scl
             bit0 = pin0
             low scl
       next j
       gosub ackn
       return


ackn : dirs = sda_in           'sda going input for recieve acknowl.
       high   scl
       if pin0 = 1 then messg   'test aknowledge bit
       low    scl
       dirs = sda_out           'sda returns to output
       low  sda
       return
```

212   *BASIC Stamp*

```
messg: serout txd,baud,(" ACKN=1 ")
       return
```

First, the initialization of the analog-to-digital converter must occur. For this initialization, the digital-to-analog converter must be written with a control byte (`PCF8591_ctrl = %01000100`), as in our first example. The four analog inputs work now as single-ended inputs, referred to as grounded. The auto increment flag is set, so the channel number will be incremented after each conversion.

In the endless loop marked with the label `loop`, the analog-to-digital converter queries start after both the start condition and the transmission of the address with four read operations. The conversion result is stored in the variable `iobyte` and will be transmitted to the host before the cycle is closed by the stop condition.

[BS1]—Digital I/O   The PCF8574 device is responsible for the enhancement of digital I/O. Notice that changes from input to output and vice versa are only possible for the whole port! The next listing (File `DIO.BAS`) shows how the two keys S1 and S2 will be queried, and depending on this result, two LEDs will be controlled. In addition to the bus lines SCL and SDA, this device has a low-active interrupt line /INT. Any changes on an input pin will be detected and will set the interrupt line to Lo. The BASIC Stamp will poll this interrupt line, making the reaction to this interrupt possible. In our programming, this polling occurs in an endless loop.

Applying a key causes an exit from the endless loop to a program part `key` responsible for querying the key. After it is known which key was pressed, the program part `set_led` drives one or both LEDs. To drive the LEDs, the port must change to output. The port of the PCF8574 device will be changed to output for a period of five seconds. In this time the LEDs concerned switch on. Applying a key at this time has no effect. To avoid short circuits, the key lines are set to Lo. After five seconds, the LEDs switch off and the program jumps into the endless loop again.

```
' -----[ Title ]----------------------------------------------------
'
' File......DIO.BAS
' Purpose...Expanded I/O with I2C-interconnection
' Author....Klaus Zahnert
' Started...07.94
' Updated...


' -----[ Program Description ]-------------------------------------
' Stamp and PCF8574 are connected with clock- and data -line
' for synchron. ser. transmission. Also connected interrupt output
```

```
' from PCF8574. External key inputs are transmitted to Stamp for
' programmed outputs to allocated LED's.

' -----[ Revision History ]------------------------------------
'
'


' -----[ Constants ]-------------------------------------------
'
    symbol sda = 0          'data            - output
    symbol scl = 1          'clock           - output
    symbol int = 2          '/INT from PCF   - input
    symbol rxd = 6          'RS232 from PC   - input
    symbol txd = 7          'RS232 to PC     - output
    symbol baud= N2400
    symbol PCF8574_wrt = %01110000 'PCF...write f. subadress 000
    symbol PCF8574_rd  = %01110001 'PCF...read  f. subadress 000
    symbol sda_in      = %10000010
    symbol sda_out     = %10000011


' -----[ Variables ]-------------------------------------------
'
        symbol iobyte    = b0      'working reg. ser.<===>par.
        symbol i         = b1      'common var., special f. loops
        symbol j         = b2
        symbol led       = b3


' -----[ Initialization ]-------------------------------------
'
        dirs = %10000011  'preset port dir. for scl,sda,rxd,txd,int
        high scl          'preset scl, sda for inact. interconn. I2C
        high sda


' -----[ Main Code ]-------------------------------------------
'
'BSAVE                           'only while syntax test without STAMP

loop:  gosub start              'initialised interrupt with dummy-input
       iobyte = PCF8574_rd      'PCF8574A read for subadresss %000
       gosub trout              'send device adress
       gosub trin               'get port input to iobyte, don't care
       gosub stop

loop1: if pin2 = 0 then key 'is there any interrupt from PCF8574?
                           'debug pin
```

```
                              '=========
        goto loop1


key:    gosub start
        iobyte = PCF8574_rd  'PCF8574 read for subadress %000
        gosub trout               'send device address
        gosub trin                'get port input to iobyte
        gosub stop
        iobyte = iobyte & %00000011   'mask for switch
        debug iobyte
        branch iobyte,(led12, led1, led2)
        ' switch          00    01    10
        ' LED                   both  green red


led1:   led = 64            'grün
        goto set_led


led2:   led = 128           'rot
        goto set_led


led12:  led = 00
        goto set_led


set_led: gosub start
        iobyte = PCF8574_wrt  'PCF8574 write for subadress %000
        gosub trout               'send device address
        iobyte = led
        gosub trout               'send led value
        pause 5000
        iobyte = 255
        gosub trout
        gosub stop
        goto loop


' -----[ Subroutines ]----------------------------------------
'
start:  low sda                   'begin transmission
        low scl
        return


stop:   high scl                  'terminates transmission
        high sda
        return
```

```
trout: dirs = sda_out
          for j = 0 to 7          'with actual databyte filled in iobyte
              pin0 = bit7          'copy act. bit from iobyte: MSB to SDA
              pulsout scl,10    'highact. clock while act. bit on SDA
              iobyte = 2 * iobyte 'next bit with shift left in iobyte
                                    'debug j,pin0
                                    'pause 200
                                    '============
          next j
          gosub ackn
          return

trin:  dirs = sda_in
          for j = 0 to 7
              iobyte=2*iobyte      'first shift is dummy here!
              high scl
              bit0 = pin0
              low scl
              next j
          gosub ackn
          return

ackn:  dirs = sda_in              'sda going input for recieve acknowl.
          high    scl
          if pin0 = 1  then messg 'test aknowledge bit
          low     scl
          dirs = sda_out           'sda returns to output
          low  sda
          return

messg: serout txd,baud,(" ACKN = 1 ")

                                    'debug pin0
                                    '==========
          return
```

These three small program examples explain the basics of the communication between BASIC Stamp and I²C bus devices. Adaptations to real applications should be easy now. All examples work as single-master systems. Remarks for multi-master systems have already been given.

For the BASIC Stamp, included debug commands (partly optional or as comments), typically allow monitoring of the data flow on the PC's screen.

All examples in this section work with the BASIC Stamp I. The differences in implementation for BASIC Stamp II are small. The enhanced memory of

BASIC Stamp II is very profitable because the I²C bus protocol requires some of the available resources.

7.2.8 [BS1]—D/A Conversion with WM8016

In the last chapter we could see that the resulting software overhead can be considerable, and the resources of the BASIC Stamps are not endless. The next example describes a very simple interface to the analog world with enough memory for an application program.

The device used for digital-to-analog conversion, WM8016 (Hughes), serves only as one example of many devices with such a simple serial interface. This device contains two independent 8-bit digital-to-analog converters with separate inputs for reference voltage and voltage-follower outputs in a 16-pin DIL or a 16-pin SO package. A simple serial four-wire interface connects the BASIC Stamp and the WM8016 device.

A peculiarity of this device is the additional on-chip implemented non-volatile memory. During power-up, the byte saved in NOVRAM will be transferred to the D/A latch. A value stored in NOVRAM is unchanged after power-up. This feature allows the use of the WM8016 for calibration purposes.

Some important technical data for the asymmetrical operation are arranged in Table 7.18.

The connection of BASIC Stamp and the WM8016 device is shown in Figure 7.52. Both reference voltage inputs will be fed from one reference voltage. Therefore, the following formula is valid for both digital-to-analog converters:

$$U_{A,B} = \frac{CODE}{256} * U_{ref} \qquad CODE = 0 \ldots 255$$

Table 7.18
Important Data of WM8016

Parameter	$V_{DD} = +5\,V \quad V_{SS} = GND$
Supply current IDD	max. 2.5 mA
Resolution	8 Bit
Total error	±1 Bit
Reference voltage	$V_{SS} \ldots V_{DD} - 2V$
Load	min. 10kΩ
Clock	max. 5 MHz
LATCHB during store	min. 10 ms

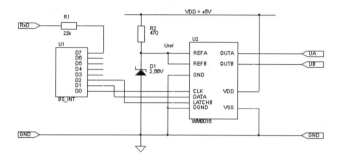

Figure 7.52
Connection of BASIC Stamp and the DA converter WM8016

When programming the WM8016 device, it is very important to take the timing into consideration. See Figure 7.53. With LATCHB = Hi and at the rising edge of the clock CLK, the data will be written in a shift register. If all data shift in, then they can be latched with the falling edge of the signal LATCHB. Note that during the falling edge of LATCHB, the clock CLK must be Hi. At the same time the coded function will be executed. Table 7.19 lists the functions.

In the following listing (File DAC.BAS), the programming of the WM8016 device for analog output is demonstrated.

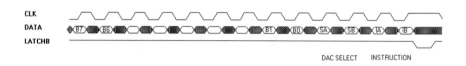

Figure 7.53
Interface Timing on the WM8016 device

Table 7.19
Functions of the WM8016 device

SA	SB	Selected DAC	IA	IB	Function
0	0	A and B	0	0	Store: DAU —> NOVRAM
0	1	A	0	1	not used
1	0	B	1	0	Load: Data —> DAU
1	1	not used	1	1	Recall: NOVRAM —> DAU

```
' -----[ Title ]-----------------------------------------------------
'
' File......    DAC.BAS
' Purpose...    Digital-Analog Conversion with WM8016 (Hughes)
' Author....    Claus Kühnel
' Started...    15.10.94
' Updated...

' -----[ Program Description ]--------------------------------------
'
' Digital-Analog Conversion with a Non-Volatile Memory DAC.
'

' -----[ Revision History ]----------------------------------------
'
' 15.10.94: Version 1.0

' -----[ Constants ]-----------------------------------------------
'
        symbol RxD    = 7           ' RS232 input
        symbol LATCHB = 2           ' Latch for DAC
        symbol DOUT   = 1           ' Serial Data for DAC
        symbol CLK    = 0           ' Serial Clock for DAC

        symbol baud   = N2400   ' Baudrate

        symbol LoadA  = %01100000   ' Load Byte to DACA
        symbol StoreA = %01000000   ' Store Byte to NOVRAM DACA

        symbol LoadB  = %10100000   ' Load Byte to DACB
        symbol StoreB = %10000000   ' Store Byte to NOVRAM DACB

' -----[ Variables ]-----------------------------------------------
'
        symbol DATA  = Pin1     ' Serial Data for DAC
        symbol Comm  = b0
        symbol Byte  = b1
        symbol Word  = w0       ' Hi-Byte = Byte; Lo-Byte = Comm
        symbol ByteA = b2
        symbol ByteB = b3
        symbol i     = b4

' -----[ Initialization ]------------------------------------------
'
        output DOUT
        low CLK
        high LATCHB
```

```
' -----[ Main Code ]-------------------------------------------------
'
start: serin RxD, baud, Comm, Byte  ' command from host
        if Comm = "A" then DACA
        if Comm = "B" then DACB
        goto start

DACA:   if ByteA = Byte then start
        ByteA = Byte
        Comm = LoadA : gosub send
        Byte = ByteA
        Comm = StoreA: gosub send
        goto start

DACB:   if ByteB = Byte then start
        ByteB = Byte
        Comm = LoadB : gosub send
        Byte = ByteB
        Comm = StoreB: gosub send
        goto start

' -----[ Subroutines ]-------------------------------------------------
'
send:   'debug %comm, %byte
        for i = 0 to 10
            Data = Bit15                ' Data = MSB
            Word = Word * 2             ' shift one position to left
            pulsout CLK, 10             ' Clock for 100 us
            'debug data
        next
        Data = Bit15
        high CLK
        'debug data
        pulsout LATCHB, 1000            ' Latch for 10 ms
        low CLK
        return
```

The program DAC.BAS is not special. The BASIC Stamp waits for two-byte commands in the form "Ax" or "Bx." The first character indicates the channel of the digital-to-analog converter. The second byte (here marked with x) defines the voltage level for analog output (CODE = 0 . . . 255).

To reduce the number of programming cycles for the internal NOVRAM, programming is only activated for changed data bytes. The cycle data transfer into latch, and the transfer into NOVRAM afterwards, is fixed programmed here.

7.2.9 Change of Circuit Parameters with NSP Devices

Changing circuit parameters outside of an integrated circuit requires special devices with controllable parameters. The common potmeter is such a device. Rotate the axle and the value of effective resistance will change. This feature makes the potmeter suitable for calibrations and adjustments of circuit parameters.

The automatization of adjustment processes by rotation of an axle would raise an amused smile today—years ago in this case a motor replaced the ghost's hand. Hint: For automatization of adjustment a motor had to rotate the axle.

NSP Family HC20xx   With HC20xx, Hughes brings a whole family of programmable analog and digital devices. A part of their features is indicated by the name. The NSP devices are **N**onvolatile **S**erial **P**rogrammable Devices. Programming is done via a synchron serial interface to the host controller. Only a few pins of the device are reserved for programming.

Architecture and peculiaritries for measurements are explained in the manuals of the manufacturer, which are helpful and very interesting. Facts important for designing the programming interface are explained only in the literature. In case of doubt, consult your distributor's material.

Figure 7.54 shows some NSP devices from the HC20xx family, with analog and switch functions. The electrical characteristics of these devices are defined by data stored in NOVRAM. After power-off, the programmed properties stay unchanged. The programming voltage is generated internally. Therefore, these devices require only one supply voltage between 3 V and 10 V DC.

Figure 7.55 describes the block scheme of all NSP devices. The different devices are distinguished through an ID code. The NV memory contents may be manipulated only if the transferred ID code is equal to the ID code of the device.

The device's specific initializations are stored in the n-Bit NV memory. If the particular device is a switch device, then the state of each switch is stored in this memory.

To avoid complex interfaces for programming and memory read operations, the two shift registers serve as a serial-to-parallel converter to the building block for the ID code compare and to the memory. Programming and memory read are therefore identical for all types of NSP devices. Finally, this is a condition for the cascading of some NSP devices.

The HC2001 device serves as an example for explaining the programming procedure. As Figure 7.54 shows, the HC2001 device is a switch device with eight independent switches. Its ID code is $D1 (= %1101 0001). Eight memory cells must be programmed, which means that eight data bits must be transferred. Figure 7.56 shows the programming procedure for the HC2001 device.

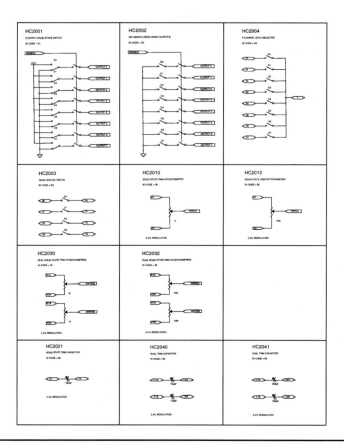

Figure 7.54
NSP devices with analog and switch functions

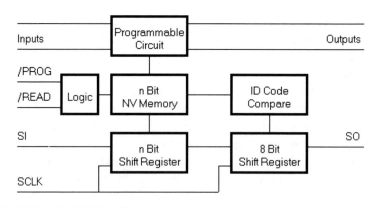

Figure 7.55
Block scheme of the NSP device

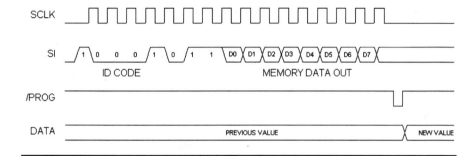

Figure 7.56
Programming the HC2001 device

The serial transmission starts with the bits of the ID code, followed by eight data bits. If all data are stored in the shift registers, then the data bits could be stored in NV memoy if the ID code compare is valid. At the end of the programming pulse, the new data are stored in the NV memory. The duration of the programming pulse must be between 1 ms and 100 ms. Because of the run-time of the BASIC Stamp, keeping the setup times is no problem. The serial clock can have any frequency under 1 MHz, and is, therefore, also not a problem.

A read-out of ID code and NV memory is important for a query of the initialization at a specific time. A read pulse with a minimum duration of 125 ns transfers the contents of the ID code register and NV memory to the shift register. After that, these data can serially read out. Figure 7.57 shows the read operation for a HC2001 device.

During read-out, the ID code data followed by NV memory data are shifted out.

Both of the discussed functions of the NSP devices are realized over three modes (Programming = SHIFT DATA + PROGRAM; Read-out = READ + SHIFT DATA). But there are two other modes, as Table 7.20 shows.

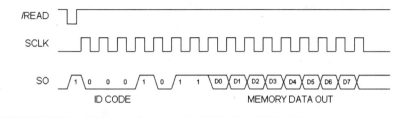

Figure 7.57
Read-out operation for the HC2001 device

Table 7.20
Truth table for NSP devices of the HC20xx family

MODE	/READ	/PROG	SCLK	SI	SO	NV DATA
STANDBY	1	1	0	X	X	VALID
READ	0	1	0	X	ID CODE	VALID
SHIFT DATA	1	1	⎍	VALID DATA IN	VALID DATA OUT	VALID
PROGRAM	1	0	X	ID CODE & DATA	X	NEW DATA
MARGIN TEST	0	⎍	0	X	⟨⟩	VALID

The most time available in the NSP device is in the STANDBY mode. The device works as defined in the programming cycle. The current consumption is reduced to a value of 10 µA maximum, typically under 1 µA. During the serial data transfer, the current consumption increases to 250 µA (SCLK = 1 MHz). This current consumption is dependent on the clock frequency. In a slow interface like that between BASIC Stamp and the NSP device, this value will no longer be reached. During the programming cycle the current consumption rises up to 1 mA during the time of the programming pulse.

The MARGIN TEST makes it possible to test the NV memory. The test will be started with a programming pulse when the input /READ is Lo. If all cells of the NV memory are okay, then the state of the serial output SO changes. The input /PROG switches back again to Hi, and the output SO will change again. If there is no toggle of the output SO, then the NV memory fails.

The performance of BASIC Stamp and NSP devices will be demonstrated in the next applications. The first example, with the HC2001 device, shows the transmission of the programming code from BASIC Stamp to the NSP device.

[BS1]—*Programmable Switch with HC2001* Controlling the segments of a seven-segment display with a programmable DIP switch, HC2001 shows the transmission of the ID code and programming pulses for a HC20xx device. Figure 7.58 shows the schema, which includes BASIC Stamp, HC2001, and an LED display HDSP-5501.

The BASIC Stamp provides an interface between the HC2001 device and a host system with an RS232 interface. The synchronous serial interface for programming the NSP device is built by Pin5 to Pin7 of the BASIC Stamp. The serial data lead to input SI of the NSP device. The output SO stays unconnected because no data can be read from the HC2001 device. The

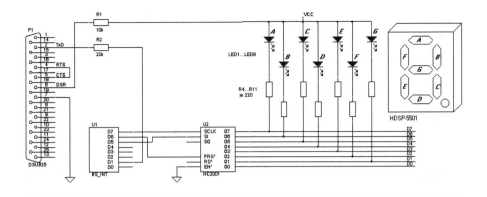

Figure 7.58
Controlling an LED display HDSP-5501

input /RD can also stay unconnected, because an internal pullup resistor generates a Hi and so avoids a Lo by accident. The input /PROG receives the programming pulse.

The data lines O7 to O1 of the HC2001 device directly control the segments of our display. According to Table 7.21, a bit pattern must be transferred into the HC2001 device. An LED segment will brighten if the HC2001 internal switch is closed. Closing a switch means a logical "0" in this bit position.

The following listing (File HC2001.BAS) shows the details of this application example.

Table 7.21
Controlling the segments of an LED display

Figure	A	B	C	D	E	F	G	Value (HEX)	Value (DEZ)
0	0	0	0	0	0	0	0	02	002
1	1	0	0	1	1	1	1	9E	158
2	0	0	1	0	0	1	0	24	036
3	0	0	0	0	1	1	0	0C	012
4	1	1	0	1	1	0	0	D8	216
5	0	1	0	0	1	0	0	48	072
6	0	1	0	0	0	0	0	40	064
7	0	0	0	1	1	1	1	1E	030
8	0	0	0	0	0	0	0	00	000
9	0	0	0	0	1	0	0	08	008

```
' -----[ Title ]-----------------------------------------------
'
' File...... HC2001.BAS
' Purpose... DIL-Switch with HC2001
' Author.... Klaus Zahnert
' Started... 13.06.94
' Updated...


' -----[ Program Description ]----------------------------------
'
' Demonstration of programming a NSP circuit with DIL-Switch HC2001
' for example.


' -----[ Revision History ]-------------------------------------
'
' 13.06.94: Version 1.0


' -----[ Constants ]--------------------------------------------
'
symbol sclk  = 7
symbol prog  = 6
symbol RxD   = 1
symbol baud = N2400      ' baudrate N2400


' -----[ Variables ]--------------------------------------------
'
symbol si    = Pin5
symbol ident = b0
symbol code  = b1
symbol memo  = w0
symbol i     = b2


' -----[ Initialization ]---------------------------------------
'
       dirs = %11100000          ' data direction of I/O pins
       high prog
       low si
       low sclk
       memo = $00D1              ' default DIL


' -----[ Main Code ]--------------------------------------------
'
loop:  for i=0 to 15
          si = bit0             ' LSB to SI
          'debug #$i,#%memo,#si,cr
```

```
              pulsout sclk,10            ' SCLK Hi for 100us
              memo = memo/2
          next i
          pulsout prog,100               ' PRG* Lo for 1ms
          serin RxD,baud,code,ident      ' read a new byte
          goto loop


   ' -------[ Subroutines ]---------------------------------------
   '
```

Four symbol definitions are used for this program. The data bytes b0 and b1 contain the ID code and the switch states of the HC2001 device. Both variables are part of the word variable memo (w0). The variable b2 with the symbol i serves as a count variable for the serial transmission of the data word w0. A test pattern will be set after the initialization of the levels on the transmission lines. If the character "8" is the test pattern, then the corresponding code is 00. The ID code for HC2001 devices is $D1. Both bytes are saved in the variable memo in this moment.

The main loop of the program now shifts the contents of the variable memo to the HC2001 device. In each step the LSB of the word variable memo will be transferred to Pin5. This bit is written in the HC2001 device with a rising edge of the clock SCLK. Before the next bit can be transferred, a right shift of the variable memo must happen. The right shift is done here by a division by two.

If all 16 bits are shifted in the HC2001 device, a programming pulse on line /PROG with a duration of 1 ms writes all bits from the shift register to the ID code, as it does to the NV memory. The HC2001 device is now programmed. The program comes to the command serin and waits for new data to write in the variable memo. If the variable memo has a new value via RS232, the whole procedure runs again.

[BS1]—Connection of Some NSP Devices   To control more than one parameter in a circuit, some NSP devices must work together. All required NSP devices must be connected together in a chain. The serial output SO of the first device will connect with the serial input SI of the second device, and so on. The /PROG, /READ, and SCLK lines of all devices build a bus controlled by the BASIC Stamp. For a band pass second order the connection of several NSP devices is demonstrated. Figure 7.59 shows the band pass circuit.

The serial input SI of device U3 is directly driven by the BASIC Stamp. Its serial output SO leads to the serial input SI of device U2. All such connected NSP devices would build a big shift register readable and writeable from BASIC Stamp.

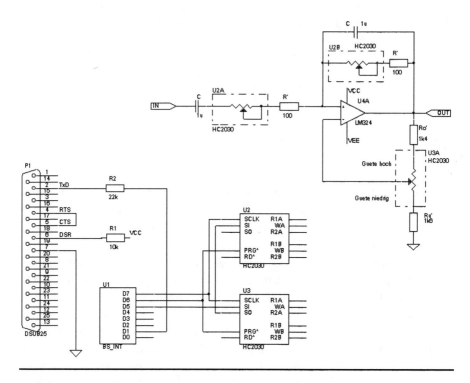

Figure 7.59
Active Wien-Bridge-Filter

After loading this shift register with the ID codes and programming bits, this data can be taken into the NV memory if all ID codes are correct. The bit sequence for programming must reflect the connection of the NSP devices.

In our Wien-Bridge-Filter, center frequency and selectivity (Q-value) are independently adjustable.

Center frequency $\quad f_0 = \dfrac{1}{2\pi RC}$

Selectivity $\quad g(\alpha) = \dfrac{f_0}{f_o - f_u} = \dfrac{1 - \alpha}{2 - 3\alpha} \quad$ with $\quad \alpha = \dfrac{R_u}{R_o + R_u}$

The ratio of the resistance divider, along with that part of the output voltage lead back to the input, defines the filter's selectivity. From the selectivity equation, the selectivity as a function of ratio α is calculated according to Table 7.22.

The steep rise of the selectivity in a small part of the voltage divider spreads the range of adjustment with the help of fixed resistors. The marked resistors R_o', R_u', and R' add to the effective part of the HC2030 device.

Table 7.22
Selectivity as a function of feedback

Ratio α	Selectivity $s(\alpha)$
0	0.5
0.2	0.57
0.4	0.75
0.62	2.7
0.65	7.0
0.655	9.85
0.6666	α

The adjustment of the center frequency requires two identical resistors and simultaneous changes of their values. The HC2030 device is qualified for this purpose. The voltage divider needs only one resistor for selectivity adjustment, so the HC2010 device is fine. For this demonstration we continued to use the HC2030. At the same time it should be demonstrated that in spite of the same ID code for two identical devices, an independent adjustment of two parameters is possible.

Figure 7.60 displays the possibilities for variations of the center frequency and selectivity, by means of PSpice simulations.

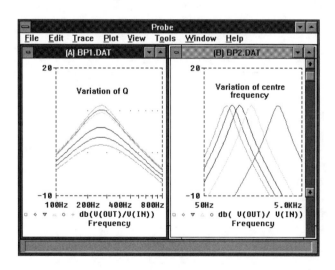

Figure 7.60
Variation of selectivity and center frequency in a controllable band pass

The following listing (File HC2030.BAS) shows an example for progamming the Wien-Bridge-Filter.

```
' -----[ Title ]------------------------------------------------------
'
' File......  HC2030.BAS
' Purpose...  Control of center frequency and Q value
' Author....  Klaus Zahnert
' Started...  20.06.94
' Updated...


' -----[ Program Description ]----------------------------------------
'
' In a RC-active bandpass filter the center frequency and the Q
' value are dependent on different resistors. HC2030 circuits build
' these resistors and so the filter performance is programmable.


' -----[ Revision History ]-------------------------------------------
'
' 20.06.94: Version 1.0


' -----[ Constants ]--------------------------------------------------
'
symbol sclk     = 7
symbol prog     = 6
symbol RxD      = 1

symbol baud     = N2400       ' baudrate N2400



' -----[ Variables ]--------------------------------------------------
'
symbol si       = pin5
symbol memo     = w0          ' double word for output conversion

symbol identg = b2            ' ID code
symbol codeg  = b3            ' Q value
symbol memog  = w1

symbol identf = b4                    ' ID code
symbol codef  = b5            ' frequency
symbol memof  = w2

symbol i        = b6
```

```
' -----[ Initialization ]-----------------------------------------
'
dirs = %11100000          ' data direction for I/O
high prog                 ' initialization of serial interface
low  si
low  sclk


' -----[ Main Code ]-----------------------------------------
'
' wait for input of Q value and frequency
loop:     serin RxD,baud,codeg,identg,codef,identf

          memo = memog
          gosub srpr       ' output Q value

          memo = memof
          gosub srpr       ' output frequency

          pulsout prog,100       ' PRG* Lo for 1ms
          goto loop

' -----[ Subroutines ]-----------------------------------------
'
srpr:     for i=0 to 15          ' shift of 16 bits
              si=bit0
              pulsout sclk,10    ' SCLK Hi for 0.1ms
              memo = memo/2
          next i
          return
```

Unlike the first example, in this instance no defined initialization is programmed. After the line initialization, the loop waits for a frequency and a selectivity value via RS232 from the PC. Four bytes providing the following information will be expected:

1. Value for selectivity
2. ID code for selectivity
3. Value for center frequency
4. ID code for center frequency

The HC2030 device will be used for both selectivity and frequency adjustment. Their ID code is 33_H. By the input of a value of $00 \ldots FF_H$, the selectivity changes in a range from 0.75 to 9.8. By the input of a value of $00 \ldots FF_H$,

Table 7.23
Programming of selectivity and center frequency

Value selectivity	ID code	Value center frequency	ID code	Programming of
00 . . . FF	33	00 . . . FF	33	Frequency and selectivity
00 . . . FF	33	XX	FF	Selectivity only
XX	FF	00 . . . FF	33	Center frequency only
XX	FF	XX	FF	Nothing

the center frequency changes in a range from 1600 Hz to 145 Hz. The spreading resistors should have tolerances of 1% or less.

The independent adjustment of selectivity and center frequency occurs by manipulation of the ID code for the parameter that is not to change. Set the ID code to a value not equal to 33_H and the parameter value will be ignored and can be set to any value. The bitstream for programming the two identical NSP devices must be built as explained in Table 7.23.

7.2.10 [BS1]—Motion Detection with a Radar Sensor

Everybody is familiar with the invisible sensors on the door of a warehouse or near a faucet or a toilet in a public restroom. Triggered by a detected motion, they execute a defined function. Most people would not guess that behind such a sensor could be a very small and compact radar sensor.

The SMX-1 module from Siemens is one such motion sensor based on modern microwave technology. Figure 7.61 shows the design of the case (25 mm × 25 mm × 14 mm).

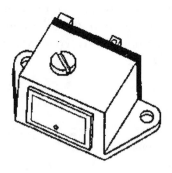

Figure 7.61
Motion sensor SMX-1

The motion sensor SMX-1 has a typical range of five meters, a transmitted frequency of 9.35 GHz and an HF output of 1 mW EIRP (**E**ffective **I**sotropic **R**adiation **P**ower). In many countries these sensors can be used without licenses, because there are no doubts of the low health risk.

The SMX-1 contains an integrated voltage regulator, which allows an input voltage between 8 and 15 V DC. The current consumption at 9 V DC is typically 25 mA. A pre-amplifier based on the double-op-amp LM358 amplifies the Doppler signal taken as voltage from a Schottky diode by a factor of 1000. The frequency response has a band pass characteristic with an upper limit of 2 Hz and a lower limit of 800 Hz. This frequency range guarantees a motion spectrum big enough for this application.

The Doppler frequency taken from the Schottky diode will be calculated as follows:

$$f_D = \frac{2 \cdot v_r \cdot f_0}{c}$$

The speed of the detected object is v_r. The microwave frequency is $f_0 = 9.35$ GHz and the light speed is a constant with $c = 3 \times 10^8$ m/s.

Using this Doppler formula, the speeds relating to the frequency limits of the band pass filter can be calculated. See Table 7.24.

The hardware required to connect the motion sensor with the BASIC Stamp is shown in Figure 7.62. The output of the integrated pre-amplifier leads over an AC coupling to the input of the op-amp U1A. The voltage gain is 10. The second op-amp U1B builds a trigger with an adjustable trigger level. The trigger output is clamped by resistor R9 and diode D1 to the supply voltage of the digital circuitry.

A Schmitt trigger sharpens the edges of the input signal for the D flip flop. This D flip flop creates a symmetrical output independent of the duty of the input signal. A symmetrical signal is required for the measurement of the period's duration. In the calculations, be sure to take into consideration the divisions by two. All components without BASIC Stamp and a speaker are parts of the printed circuit board on SMX-1.

Evaluation of a motion signal and classification in two speed ranges are demonstrated with the following program RADAR.BAS.

Table 7.24
Measured speed vs. Doppler frequency

Doppler frequency	Speed	
2 Hz	0.032 m/s	0.115 km/h
750 Hz	12.03 m/s	43.43 km/h

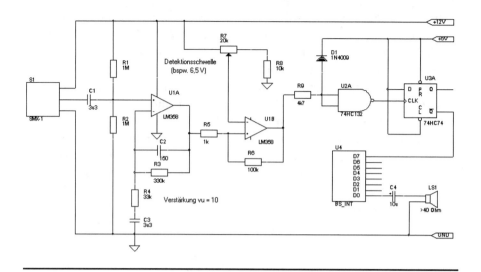

Figure 7.62
Connecting the motion sensor SMX-1 to BASIC Stamp I

```
' -----[ Title ]----------------------------------------------------
'
' File......  RADAR.BAS
' Purpose...  Motion detection with motion sensor Siemens SMX-1
' Author....  Claus Kühnel
' Started...  10.10.94
' Updated...


' -----[ Program Description ]--------------------------------------
'
' Demonstration of a simple motion detection based on radar doppler
' effect.


' -----[ Revision History ]----------------------------------------
'
' 10.10.94: Version 1.0


' -----[ Constants ]------------------------------------------------
'
        symbol sensor   = 7         ' pin for sensor input
        symbol speaker  = 0         ' pin for speaker output


' -----[ Variables ]------------------------------------------------
'
```

```
        symbol value = w0


        symbol lo_range  = w1
        symbol mid_range = w2
        symbol hi_range  = w3


' -----[ Initialization ]-----------------------------------------
'
        dirs= %10000000          ' initialization of input & output
        value = 0


        lo_range  = $0000        ' quick motion range
        mid_range = $8000        ' ------------------
        hi_range  = $ffff        ' slow  motion range


' -----[ Main Code ]-----------------------------------------------
'
start: pulsin sensor, 1, value      ' puls detection
        pause 500            ' classification in slow and quick motion
        if value >  lo_range  and value < mid_range then alarm1
        if value >= mid_range and value < hi_range  then alarm2
        high speaker
        goto start               ' repeat endless


alarm1:sound speaker,(40,10,80,10)                 ' quick motion
        value = 0
        goto start


alarm2:sound speaker,(80,10,120,10,80,10,120,10)   ' slow motion
        value = 0
        goto start


' -----[ Subroutines ]---------------------------------------------
'
```

With the instruction pulsin sensor,1,value, the BASIC Stamp I is waiting for a rising edge at Pin7. After this rising edge the time to the falling edge
of this pulse will be measured with a resolution of 10 µs. All measured speeds
will be placed in two classes, according to their defined constants. Table 7.25
shows the values for these speed classes. In this case, a slow motion generates
another order of tones as a quick motion.

Table 7.25
Classification of motion events

	Quick motion	*Slow motion*
Range	1 . . . 7FFF$_H$	8000$_H$ . . . FFFE$_H$
Doppler period	0.01 ms . . . 327.67 ms	327.68 ms . . . 655.34 ms
Doppler frequency	100 kHz . . . 3.05 Hz	3.05 Hz . . . 1.52 Hz
Speed	(3208 m/s) . . . 0.0978 m/s	0.0978 m/s . . . 0.0488 m/s
	(11,549 km/h) . . . 0.352 km/h	0.352 km/h . . . 0.176 km/h

7.2.11 [BS1]—Distance Measurement with IR Triangulation

If the simple Yes/No decision of a light barrier or a proximation circuit as a way of monitoring is not sufficient, then the distance measurement can deliver additional information. Simple systems for distance measurement should be not more expensive than light barrier systems. In this section, Sharp's position-sensitive photo detector GP2DO2 will be described, as it relates to function and application.

The sensor GP2DO2 contains an infrared transmitting diode, a position-sensitive photo detector, some optics, and a signal-processing unit in a compact casing of $29 \times 14 \times 14.4$ mm³. Only four pins give the electrical contact to the sensor. Two of them are power supply pins and the other two build a simple serial interface for data handling during the measurement. Figure 7.63 shows Sharp's sensor GP2DO2.

On the left side of the sensor's case, there is an IR LED, and on the right side is a position-sensitive photo detector behind an optical lens. The method of distance measurement, shown in Figure 7.64, is very simple. The IR LED transmits a bundled beam to the object plane to be measured. The photo detector receives a reflected beam. The angle of the received beam depends on the distance of the object plane. The situation for two different object

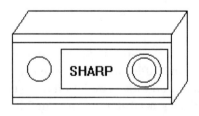

Figure 7.63
Sensor GP2DO2

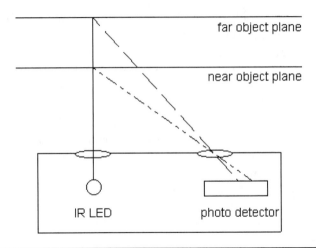

Figure 7.64
Distance measurement by triangulation

planes is displayed in Figure 7.64. If the photo detector can process a position-sensitive input signal, then information on the distance of an object is given.

Photo diodes with a big light-sensitive area have position sensitivity on principle. Based on this physical effect, position-sensitive photo detectors were developed. Figure 7.65 displays the principle of a position-sensitive photo diode. The vertical structure of the position-sensitive photo detector is identical to the structure of general pin diodes. At the bottom is an n-conductive substrate layer, then an isolation layer. Embedded in this isolation layer is the p-conductive layer irradiated by IR. The contacts to the p-layer (anode) are made on the left and the right side.

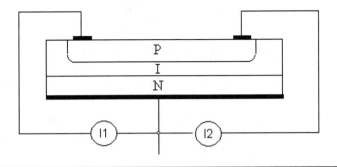

Figure 7.65
Structure of a position-sensitive photo diode

If there is a spot irradiation in the center of the p-layer, then both currents I1 and I2 will have the same value. If the spot irradiation goes to the left, the current I1 will grow and the current I2 will drop with the same value. The difference between the currents I1 and I2 describes the location of a spot irradiation on the light-sensitive surface (between the contacts) of the position-sensitive photo diode.

To process the measured distance, the signal processing unit has to exploit the difference in currents I1 and I2. After an analog-to-digital conversion, the result of the measurement is ready for serial transmission to a microcontroller or other processing unit.

For the processing of optical signals, comparable conditions are valid:

- Daylight (or the light of the environment, in more general terms) may not influence the measuring procedure.
- Distance and reflection of objects strongly affect their values.
- The very high dynamic range of the measuring signal requires signal compression for working with normal voltages of the power supply.

Disregarding special circuit techniques in bipolar integrated circuit technology (like I^2L) that convert an irradiation directly into a digital signal, we can easily find circuits based on logarithmic current-to-voltage converters. For logarithmic current-to-voltage converters, there are excellent technological solutions with very good performance.

Figure 7.66 shows the principle of a circuit for position-sensitive current-to-voltage conversion. The position-sensitive photo detector is described through the photo diodes FD1 and FD2. Both anodes are connected with identical circuits for current-to-voltage conversion. The diodes in the op-amp's feedback cause a logarithmic behavior in the current-to-voltage conversion.

To achieve the required accuracy, the diodes are built by bipolar transistors with a grounded base. In a good approach the following equation is valid:

$$I_C \cong I_0 \cdot \exp\left(\frac{V_{BE}}{V_T}\right)$$

V_T describes the temperature voltage proportional to the absolute temperature.

$$V_T = \frac{kT}{e} \quad \text{with } k = \text{Boltzmann's constant and } e = \text{elementary charge}$$

I_0 describes the very strong temperature-dependent saturation current of the bipolar transistor. For the output voltage of the logarithmic current-to-voltage converter, the following equation is valid:

$$V_0 = -V_T \cdot \ln\left(\frac{I_C}{I_0}\right)$$

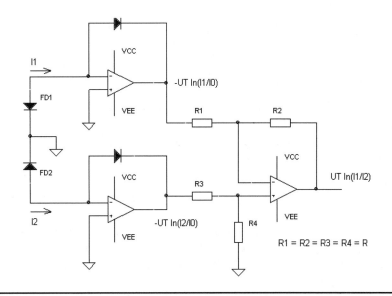

Figure 7.66
Circuit for position-sensitive current-to-voltage conversion

The collector current I_C is identical to the current I_1 in the upper circuit part of Figure 7.66 and to I_2 in the lower part. The third op-amp processes the difference in the two output voltages. There is only one condition for accuracy—the resistor values have to be equal.

The output signal of the third op-amp is described by the following equation:

$$V_0 = V_T \cdot \ln\left(\frac{I_1}{I_2}\right)$$

The influence of the saturation current I_0 is canceled. Only the influence of the temperature voltage remains. Because this influence is linear it can easily be eliminated. Changes of intensity (daylight) will be suppressed because they affect both currents. For further digital processing, the output voltage must convert to digital.

Some statements on the inner processes for distance measurement can be derived from the timing diagram in Figure 7.67. The two lines V_{in} and V_{out} are available for controlling the measurement and data handling. V_{in} must be connected to an open-drain output. V_{out} gives an output compatible to CMOS and/or TTL.

To start the distance measurement, the control line V_{in} must drop to Lo for a minimum time of 70 ms. In this time the IR LED transmits sixteen pulses in

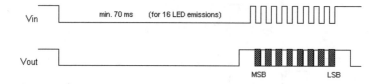

Figure 7.67
Timing diagram for measurement and data handling

the direction of the object to be measured. These sixteen measurements allow the calculation of the mean value for reducing possible errors.

After this measuring phase, the result can be called from the sensor. In this phase, the control line V_{in} has the function of a synchronous clock input. Starting with the most significant bit (MSB) the eight data bits are available at V_{out}.

Controlling the sensor requires few resources of the connected BASIC Stamp (BS1-IC or BS2-IC). Figure 7.68 shows the minimum system for distance measurement and display. Additional I/O pins can serve as status outputs. All three units of the whole circuit are powered by the same supply voltage of +5 V DC. I/O line D0 transmits the data to a serial–controlled LCD, the LCD Serial Backpack from Scott Edwards offered by Parallax, Inc. I/O line D1 drives the LED to visualize the start of measurement. The input V_{in} of the sensor GP2DO2 is connected to the FET Q1, to fulfill the requirements for an open-drain driver. The serial result of distance measurement will be the readout over I/O line D3.

The following listing (File `GP2DO2.BAS`) shows the control program for cyclic distance measurements in PBASIC for BASIC Stamp I. Changes for BASIC Stamp II are very simple.

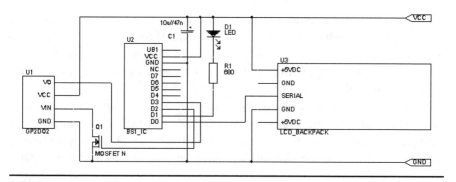

Figure 7.68
Minimum system for distance measurement and display

```
' -----[ Title ]----------------------------------------------------
'
' File......   GP2DO2.BAS
' Purpose...   Test of Distance Measuring Sensor Sharp GP2DO2
' Author....   Claus Kühnel
' Started...   18.03.95
' Updated...


' -----[ Program Description ]-------------------------------------
'
' The distance measuring sensor Sharp GP2DO2 Type 1 measures
' distances between 10 cm and 80 cm.
' The result of 16 measurements is given serial as a count byte.
' The BASIC Stamp controls the starts the measurements, reads the
' result and send it serial to an LCD Serial Backpack Rev. 3.


' -----[ Revision History ]----------------------------------------
'
' 18.03.95: Version 1.0


' -----[ Constants ]-----------------------------------------------
'
        symbol LCD    = 0        .' serial output to LCD
        symbol LED    = 1        ' LED control
        symbol Vin    = 2        ' sensor control
        symbol baud   = N2400    ' Baudrate

' -----[ Variables ]-----------------------------------------------
'
        symbol count = b0        ' result of measurement
        symbol i     = b1        ' loop variable
        symbol Vo    = pin3      ' sensor data


' -----[ Initialization ]------------------------------------------
'
init: low Vin
      high LED
      count = 0
      gosub LCD_clear

' -----[ Main Code ]-----------------------------------------------
'
start:pulsout LED,20000
      gosub sensor_control
```

```
        gosub LCD_print
        pause 1000
        goto start


' -----[ Subroutines ]----------------------------------------
'
sensor_control:
        'debug cls
        high Vin
        pause 70
        count = 0
        for i = 0 to 7
            pulsout Vin,10
            count = count * 2 + Vo
            'debug %count
        next i
        low Vin
        'debug count
        return

LCD_clear:
        serout LCD,baud,(254,1,254)
        serout LCD,baud,("Count:")
        return


LCD_print:
        serout LCD,baud,(254,136,254,#count," ")
        return
```

At the start of each cycle, the LED flashes. In the subroutine `sensor_con-`
`trol` the measurement starts, followed by the serial readout of the result. After
the readout, the result is stored in the variable `Count`. The subroutine
`LCD_print` works like a "print-at" instruction for the LCD and writes the
result as a decimal number to a defined position. After a waiting period of one
second, the whole procedure will be repeated.

Table 7.26 shows test results for a reflector built from white copy paper. Figure 7.69 shows a graphical representation of these measuring results.

Table 7.26
Counts in displays vs. distance

Distance in cm.	10	20	30	40	50	60	70	>80
Counts in display	147	91	71	55	47	44	42	40

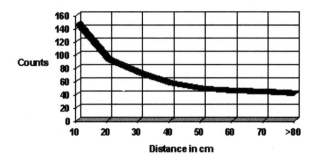

Figure 7.69
Distance measurement

An important parameter describing the sensitivity of the sensor is the output-distance characteristic ΔVo. This parameter will be represented by an output change as a result of a distance change from 80 cm to 20 cm. The values of ΔVo are between 45 and 65 (typically 55). The measured value here was 51 counts, a little smaller than the typical value. The measuring distance range is limited to distances from 10 cm to 80 cm. Distances smaller then 10 cm give angles for the received beam that do not work.

7.2.12 [BS2]—Scale Based on Frequency Output

The command freqout is one of the special commands of BASIC Stamp II. Of course it is impossible to get analog values from a BASIC Stamp. It has no digital-to-analog converting device, and neither does the BS2 carrier board.

To realize analog voltages at a BASIC Stamp II pin, we use a trick already familiar from the command PWM. Periodical pulses with variable pulse lengths are made. After filtering by a low-pass, one gets the average value of the voltage. Small pulses make a low voltage, longer pulses a higher voltage, limited by the pulse's amplitude. The amplitude is the value of the logical Hi state, approximately +5 V. On the other side, a very small pulse lets the voltage go to zero. A sine wave is realized with a periodical change of pulse length, depending on the actual amplitude of the sine wave. The program shown in the next listing (File SCALE1.BS2) provides tone output like a scale for some octaves.

```
' -----[ Title ]-------------------------------------------
' File...... scale1.BS2
' Purpose... output scale of tone
' Author.... Klaus Zahnert
' Started... 96.1.2
```

```
' Updated...


' -----[ Program Description ]-------------------------------------
'Loop contains output of ton scale of some octaves.
'Each count of loop is marked by led for possible break with key


' -----[ Revision History ]----------------------------------------


' -----[ Constants ]-----------------------------------------------
      spin   con     6
      time   con 1000            'setting duration tone-output /ms
      octv   con     5           'setting number of generated octaves
                                  ' 1.....5


' -----[ Variables ]-----------------------------------------------
'

      n    var byte              'count for  loop tone
      o    var byte              'count of loop octave
      ff   var byte              'multiplier frequence
      f    var word              'frequence /Hertz
      t    var byte              'tone index
      key var IN14               'key input


' -----[ Initialization ]------------------------------------------
'no


' -----[ Main Code ]-----------------------------------------------
'

Start: low 15
        debug " ", 10,13          'space line for return full loop
        For o = 1 to octv
        lookup o,[0,1,2,4,8,16],ff 'frequ.- factoring for each octave
        For n = 0 to 7
        lookup n,[261,294,330,370,392,440,494,523],f    'scale tone
        lookup n,["c","d","e","f","g","a","b","c"],t     'mark tone
        debug "f = ",dec4 ff*f," Hz   tone =  ",t,10,13 'out screen
        freqout spin,time,ff*f                          'tone output
        next
        debug " ",10,13           'space line in octave
        pause 1000                'delay
        next
        high 15                   'LED on 1 sec
        pause 1000
        If key = 0 then raus      '..with possible
        goto start                ' break of loop
```

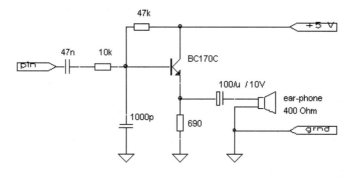

Figure 7.70
Ear phone amplifier

```
raus : end

' -----[ Subroutines ]-----------------------------------------
'no
```

You may hear the scale by connecting a speaker or ear phone to Pin6. To give some power to a speaker with included low-pass and impedance transformation we used a little circuit between the speaker and BASIC Stamp II Pin6, as shown in Figure 7.70.

The permanently-installed LED on Pin15 and the key on Pin14 are used to break the loop from its continous working. Duration, tone output, and the count of octaves may vary according to the constants.

Keep the lookup statement twice in the program. The first time, it is like a table of programmed frequencies with one musical tone name. The second time, it is "factor generating" for multiplying frequencies in certain octaves. In this way, each tone of an octave, followed by the first, has the factor value from the above decade.

7.2.13 [BS2]—Generate and Recognize DTMF Tones

In the same way that telephones went from using mechanical to electronic components in the last century, in more recent years the method to transmit dialed messages has changed from pulse to tone. Each dialed digit is transformed to a combination of two frequencies of around 1 kHz as a characteristic tone pair. The specific ratio is chosen to avoid complications with harmonics and signals of human speaking. Table 7.27 shows the frequencies for each dialed character.

Table 7.27
Functional decoding CS8870

F_{low}	F_{high}	KEY	Q4	Q3	Q2	Q1
697	1209	1	0	0	0	1
697	1336	2	0	0	1	1
697	1477	3	0	0	1	1
770	1209	4	0	1	0	0
770	1336	5	0	1	0	1
770	1477	6	0	1	1	0
852	1209	7	0	1	1	1
852	1336	8	1	0	0	0
852	1477	9	1	0	0	1
941	1336	0	1	0	1	0
941	1209	*	1	0	1	1
941	1477	#	1	1	0	1
697	1633	A	1	1	0	1
770	1633	B	1	1	1	0
852	1633	C	1	1	1	1
941	1633	D	0	0	0	0

DTMF tones are not only usable for phones; devices for remote control often need cipher signals on analog channels. For example, an amateur radio station gives coded signals to a remote relay station to posit antenna direction for communication with local stations.

Here we have the statement DTMFOUT, like the statement FREQOUT in BASIC Stamp I. To demonstrate tone generation, we use a speaker on Pin6 again, connected in the same way as shown in the last section for the ear phone amplifier.

The program is very short. A continuous loop displays acoustical and visual signals of the digits 0 . . . 9. Included is the possibility of keybreaking with Pin14.

```
start: for n=0 to 9
          DTMFout 6,500,500,[n]
          debug dec n,10,13
       next
brk1 : if in14=0 then raus
       goto start
raus : end
```

Each dial 0 . . . 9 is output on Pin6 with a tone duration of 500 ms, followed by a delay of 500 ms.

Reversing back from tone to dial is not implemented in the BASIC Stamps, because it is more difficult than tone generation. Several digital filters and special hardwired components would be needed to recognize and interpret the tone combinations. Some semiconductor firms produce a special type of integrated circuit for this purpose, reliably and cheaply. One of them—a "classic model"— is the 8870-IC. We have used the MITEL MT8870. Also usable is the CRYSTAL CS8870, and there are others. For our application only the most remarkable features will be shown. See supplemental documentation from the manufacturer.

Table 7.27 shows the frequency pairs on the input to assigned keys and the pattern Q4 . . . Q1. The timing for important signals is shown in Figure 7.71.

To use the scanning dial with BASIC Stamp II, we connect the MT8870, as Figure 7.72 shows.

Outputs Q4 . . . Q1 of MT8870 are connected to Pin3 . . . Pin0 of BASIC Stamp II for transmission of the decoded dial signals. The tone input for MT8870 is on Pin2. The gain for the dial tone is controlled by the ratio of resistors R1 and R2. R2 is the feedback of the inner op-amp. We set the resistor ratio to 100 kΩ/470 kΩ for an input-voltage of approximately 1 V$_{rms}$. We got the signals from the ear socket of a dictaphone. For timing, R = 330 kΩ and C = 0.1 µF is needed. Ceramic capacitors near the MT8870 are efficient to prevent spikes in the power supply. A connection between Pin1 and Pin4 serves as a reference signal to the inner op-amp. TOE on Pin10 is permanently connected with Hi. If Lo, Q1 . . . Q4 switch in tri-state condition. A very important signal is StD ("Delayed steering Output"). It marks the end of a conversation with stable values Q1 . . . Q4, after recognizing a dial tone.

The pulse diagram in Figure 7.71 lets us see how to get real values for dialing. It is only a simplified scheme to understand the connection and the program, to get success in dial display with BASIC Stamp II.

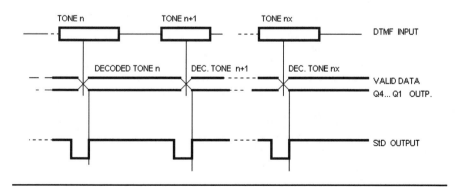

Figure 7.71
Simplified pulse diagram MT8870 for DTMF input, output Q4 . . . Q1, and StD

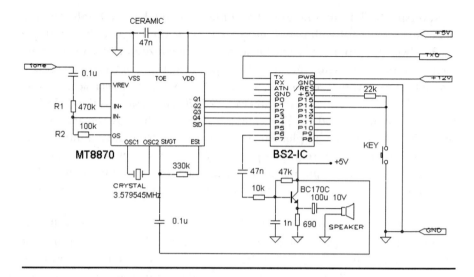

Figure 7.72
Connecting MT8870 to BASIC Stamp II

The speaker on Pin6 of BASIC Stamp II is the same as that used in this section, Scale Based on Frequency Output. We caught the DTMF signals with a dictaphone held by hand near by its membrane to save the tape and to replay as described. In the same way we got phone line signals with a coupled dictaphone. The program for evaluating the dial tones is listed as follows (File MT8870.BS2).

```
' -----[ Title ]-------------------------------------------------
' File......MT8870.Bs2
' Purpose...Receive DTMF-dial-tone,display corresponded digits
' Author....Klaus Zahnert
' Started.96.1.20
' Updated...


' -----[ Program Description ]-----------------------------------
' DTMF-Decoder MT8870 started with input each dial-tone. Output is
' marked by status-signal StD = high. This signal startet
' transmission BCD-output, generated from decoded tone- signal, to
' input - pins BS2
'


' -----[ Revision History ]-------------------------------------
```

```
' -----[ Constants ]-------------------------------------------------
'
          spin   con 7                        'serial output
          baud   con 84+$4000                 'baudrate 9600


' -----[ Variables ]-------------------------------------------------
'
          BCD    var   byte
          dial   var   byte
          StD    var   IN4
          key    var   IN14


' -----[ Initialization ]--------------------------------------------
'
  DIRL = %01000000


' -----[ Main Code ]-------------------------------------------------
'
start:
        debug cr,"phone = "   ' announce dial scan
        serout spin,baud,["phone = "]
dgt:  if StD=1 then scan    ' wait for decoded dial or key for end
dg1:  if key=1 then dgt
        if key=0 then fin
scan: BCD = INL & $0F       ' eleminate signal Q1...Q4
        lookup BCD,[255,49,50,51,52,53,54,55,56,57,48,42,35],dial
                          ' assign BCD ==>ASCII of 1,2,....,8,9,0,*,#
        debug dial         ' display digit
        serout spin,baud,[dial,13,10]

hold: if StD=1 then hold ' hold while actual digit
        goto dgt              ' jump to possible next digit or keybreak

fin:  debug 13,10,"aborted dial scan",13,10
        end                  ' message after keybreaking with end progr.


' -----[ Subroutines ]-----------------------------------------------
'no
```

For output, we use the command debug, or a serial output to a monitor. We
decode only the digits 0 . . . . 9 , *, #. It is easy to enhance the lookup table for
the characters A . . . . D, if needed.

   The program is controlled by polling StD input. The first loop, marked
by dgt and dg1 is left when StD goes Hi to work with Qn data. Possible

keybreaking is included again. So there will be only one display for one conversation of the MT8870, there is the loop `hold` to recognize when the `StD` signal goes low for the start of the next conversation.

7.2.14 [BS1]—Light Measurements with TSL230 Sensor

In the following application example, a BASIC Stamp I is used to convert a frequency-modulated output of a light sensor in a standard RS232 signal. Because there is a standard communication channel, an application program, running on a PC and written in Visual BASIC for Windows 3.0, can communicate with the light sensor itself.

Light Measurement Based on Light-to-Frequency Conversion   The use of silicon photo diodes to measure visible irradiation is quite familiar. Their detectable spectral range extends from about 300 nm up to 1100 nm. Corrections to the spectral sensitivity of the human eye must be made by optical filters. These corrections are required for all measurements that presuppose the spectral sensitivity of the human eye. The exposure measurement in photo cameras is a typical example.

The short circuit current of a silicon photo diode is proportional to the incidental irradiation, and is also nearly independent of the temperature. It is detrimental that in general only small currents are generated. So the circuits in the analog part of a device for opto-electrical conversion can become expensive.

A perceptible reduction of the required expense will be reached, if the photo diode and the circuit for amplification are integrated on the same chip or the same device.

Texas Instruments offers the TSL230, a programmable light-to-frequency converter in an 8-pin clear plastic dual-in-line package. A silicon photo diode and a current-to-frequency converter are integrated.

In the case of maximum sensitivity, an irradiation of 450 µW/cm² ($\lambda p = 660$ nm) typically gets an output frequency of 1 MHz. The light-sensitive area of this photo diode is typically 1 mm².

The photo diode is configurable, which means its sensitivity can vary by factors of 10 and/or 100. Additional division of output frequency makes adjustments to concrete conditions possible. Figure 7.73 shows the elements of the TSL230 device in a block diagram.

Besides the control inputs S3 to S0, there is a lo-active output-enable input. This output-enable can switch the output of the device in a tri-state condition. So outputs of several devices can be connected to a common output line. The conditions listed in Table 7.28 are valid for programming. The first line in the table is special, because certain issues must be considered for this combination

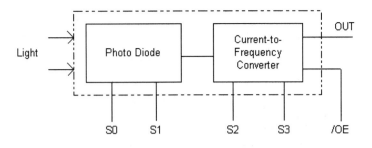

Fig. 7.73
Block diagram TSL230

Table 7.28
Programming the TSL230 device

S3	S2	Frequency Scaling	S1	S0	Sensitivity
L	L	f	L	L	Power down
L	H	f/2	L	H	1x
H	L	f/10	H	L	10x
H	H	f/100	H	H	100x

of parameters. Frequency division causes a symmetrical output (duty = 0.5). In the case of S3 = S2 = L, the duration of an output pulse extends from 125 ns to 500 ns, so symmetry is not certain.

In the case of S1 = S0 = L, the device switches to the power-down mode. The supply current is reduced from 2 μA to a maximum of 10 μA.

BASIC Stamp as an RS232 Interface   The TSL230 is programmed through the control inputs S3 to S0. To explore the irradiation, the output frequency of TSL230 must be measured. Interfacing the TSL230 device to a host computer makes it possible to do all other tasks from the host side. A simple solution results, if this interface is built by a BASIC Stamp. Figure 7.74 shows the entire interface circuit.

The interface to a host computer is built by a simple 3-wire connection, according to the RS232 standard. The source program (File `TSL230.BAS`) listed below explains the work of this interface circuit.

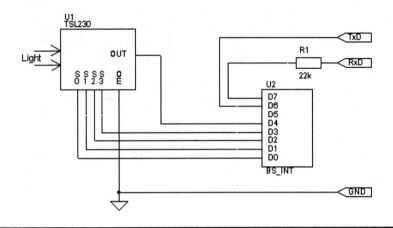

Figure 7.74
RS232 interface to TSL230 device

```
' -----[ Title ]-----------------------------------------------------
'
' File......   TSL230.BAS
' Purpose...   RS232-Interface for TSL230
' Author....   Claus Kühnel
' Started...   29.10.94
' Updated...

' -----[ Program Description ]----------------------------------------
'
' The programmable light-to-frequency converter TLS230 (TI) is
' interfaced to a host controller by a simple RS232.
' A command sets the inputs S3 to S0 of TLS230. After this setup
' the Stamp sends the measured value to the host.

' -----[ Revision History ]------------------------------------------
'
' 22.11.94: Version 1.0

' -----[ Constants ]-------------------------------------------------
'
            symbol RxD     = 7
            symbol TxD     = 6
            symbol LED     = 5
            symbol TSL230  = 4
            symbol baud    = N2400
```

```
'  -----[ Variables ]----------------------------------------------
'
            symbol command = b0
            symbol value   = b1
            symbol period  = w1
            symbol ss      = b4


'  -----[ Initialization ]-----------------------------------------
'
            dirs = %01001111
            pins = %00000101


'  -----[ Main Code ]----------------------------------------------
'
            period = 0
start:      serin RxD, baud, command, value
            if command <> "S" then dummy
            high LED
            lookdown value,("D","E","F"),ss
            pins = 14 + ss


            pulsin TSL230,0,period
dummy:      serout TxD, baud, (b3,b2)
            period = 0
            goto start


'  -----[ Subroutines ]--------------------------------------------
'
```

After some definitions and initializations, the program starts at the label start, waiting for two characters from the serial interface. The host sends the required commands to the BASIC Stamp I. The setup of the control lines S3 to S0 depends on the second character.

The Lo-time of TSL230 output pulses is measured with the command pulsin. To calculate the output frequency from this Lo-time, the pulse sequence must be symmetrical. Therefore, use of the hatched condition in the state table is forbidden.

The result from the command pulsin is saved in the 16-bit variable period. The resolution is 10 µs. After determining the half-period duration, the result can be transferred to the host. The BASIC Stamp I sends only the two resulting bytes to the host. After sending the conversion result, the program jumps back to wait for a new command.

Handling the PC's COM Port For including the serial data in an application program under Microsoft Windows™, Visual BASIC for Windows in Version 3.0 provides excellent support for program design without a long training period.

An important point of the application program is the handling of serial data. In general there are two possibilities—interrupt driven or polling. Both possibilities are supported by Visual BASIC V. 3.0.

Because the BASIC Stamp I sends its result a short time after receiving its setup command, the polling variant was used here. The application program described here is a simple monitor program called TSL230 Light Monitor. Its tasks are

- Initialization of a COM port.
- Initialization of a timer for requesting the sensor device and the PC clock for time and date.
- Setup of the sensitivity of the sensor device.
- Driving some displays.
- Registration of max-min-value as a calculation example.

Figure 7.75 shows the user interface of the TSL230 Light Monitor. In the bottom left area of the screen are the option buttons. Regarding the programming table of the TSL230 device, there are three options. Every second, the timer sends a setup command to the sensor device, and the answer from this device is the measured value proportional to the light. If this value is higher than the momentary maximum or lower than the momentary minimum, then this value will be corrected and stamped with the time and date.

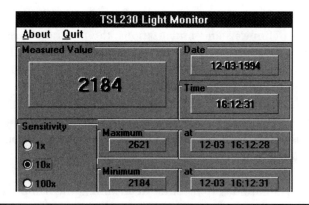

Figure 7.75
User interface of TSL230 Light Monitor

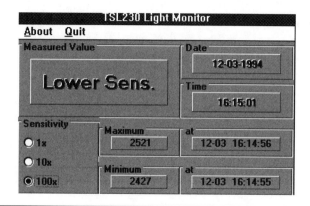

Figure 7.76
Sensitivity is too high—Hint to lower the sensitivity.

Under some conditions the measured values may be too high, because the adjusted sensitivity does not conform. In these cases no value is shown, and a hint for changing the sensitivity is displayed. Figure 7.76 shows the case of sensitivity that is too high.

To give an impression of the data handling, the next two listings provide subroutines of the monitor program.

```
Sub Form_Load ()
    comm1.CommPort = 2
    comm1.Settings = "2400,N,8,1"
    comm1.InputLen = 0
    comm1.PortOpen = True
    Option3D2.Value = True: param$ = "SE"      ' Sensitivity 10x
    max = 0
    min = 65535
End Sub
```

After the start of the program, a load event occurs. The user interface will be displayed and the initialization of the COM port will be done. To simplify, here COM2 is used with no possibility of change. The other communication parameters are given by the BASIC Stamp.

The initial value of sensitivity is 10x. The setup command for the TSL230 device is "SE" in this case, each second a timer event occurs. The subroutine `Timer1_Timer` contains all activities, which run automatically.

```
Sub Timer1_Timer ()
    Dim dummy, mess$
```

```
DisplDate.Caption = Date$
DisplTime.Caption = Time$

comm1.Output = param$
Do
    DoEvents
Loop Until comm1.InBufferCount >= 2
mess$ = comm1.Input

Value = Asc(Left$(mess$, 1)) * 256
Value = Value + Asc(Right$(mess$, 1))
If Value = 0 Then
    DisplMValue.Caption = "Lower Sens."
Else
    Value = 65535 / Value
    DisplMValue.Caption = Str$(Value)
    maxmin
End If
End Sub
```

7.2.15 [BS2]—Serial Reading from Phone Card

Every day, many people use chip-based phone cards. After the account goes down to zero, the cards seem to be good for nothing, although some people collect these cards for their printed pictures. Yet each card provides specific information on its EEPROM: a five-digit serial number. You can use this number like a "hardware password" to get security for entering doors or starting machines, computers, and so on.

For reading the information saved in the phone card we will use the synchronous serial data transmission of the BASIC Stamp. For BASIC Stamp I we have to program the shift operation with PBASIC statements. For BASIC Stamp II we use the comfortable shiftin statement.

First, let us look at the construction of the phone cards: It is based on ISO 7816. There is a static EEPROM with 128 bytes. The serial synchronous transmission is made by clock and data line only. With power and reset we have five wires to connect the card. Figure 7.77 shows the connectors on the card. You can see the timing of the described signals RES, CLK, and DATA in Figure 7.78.

The read operation starts with a CLK pulse while RES is Hi. With RES = Lo, each CLK pulse delivers a one bit or DATA line. There are 128 bits clocking out in the same way. So you get fifteen bytes, all the information marking the card and counting down the credit remaining with every suc-

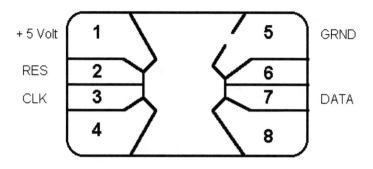

Figure 7.77
Phone card connectors

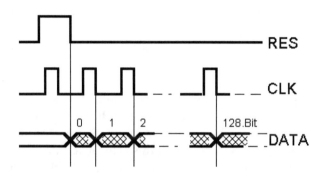

Figure 7.78
Timing for data transmission

cessful phone process. The serial number of a German phone card has five digits, each a nibble of four bits. The position in the chain of serial transmission is bit 43, starting with the LSB of the lowest digit nibble. Table 7.29 shows an example.

The serial number 2 9 3 5 2 is **bold** from the fifth to seventh byte. We are not concerned with all the other information on that bytes. But you can take

Table 7.29
Bit stream, red out from phone card

Byte	00	01	02	03	04	05	06	07	08	09	0A	0B	0C	0D	0E	0F
cont. hex	F4	4F	FF	25	74	**B2**	**53**	**92**	00	00	80	FE	F8	FF	FF	FF

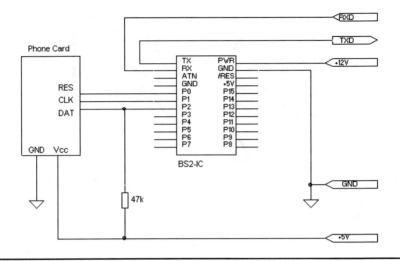

Figure 7.79
Phone card reader hardware

all the information on the card, or a specific feature, for example, the account of possible phone calls. Unfortunately with one shiftin statement you get only a maximum of sixteen bits, the contents of a word. So we need one shiftin with four nibbles, the digits 2 - 5 - 3 - 9 in this order. The next shiftin carries the most significant digit 2 with a one-nibble transport. To start the shiftin process at the proper position in the bit chain, we skip the unnecessary bits with clock pulses. The following listing (File CHIPCRD1.BS2) shows the needed program. The hardware of the phone card reader is shown in Figure 7.79.

The listed program shows the solution for serial data transmission once more in two ways. You can alternate the shiftin operation for sixteen and four bits in accordance with the subprograms again. So you have a way to implement for the BASIC Stamp I, too.

```
' -----[read serial number of chipcard]-----------------

' File..... chipcrd1.bs2
' Purpose.. use blank phone-chipcard's serial-number
'           to read as special password
' Author... Klaus Zahnert
' Started.. 95.12.16
' Updated..

' -----[Program description]-------------------------
'       read serial number of phone-chip-card ISO 7816
```

```
'         to PC with manual programmed ser. data-transm.
' -----[Revision History]--------------------------------
'
' -----[Constants]---------------------------------------
RES     con 0               'output: Reset-line
CLK     con 1               'output: Clock -line
DAT     con 2               'input : ser. Data - line
TIME    con 10              'Clock-on-time
POS     con 43              'Adressposition bit 43


'------[Variables]---------------------------------------
MEM16   var word    'four lower position serial number
MEM4    var nib     'one  highest pos. serial number
bit_nr var byte     'index bitposition


'------[Initialisation]----------------------------------
        dirl=$03    'CLK,RES are outputs
        low RES
        low CLK
        mem16 = 0
        mem4  = 0


'------[Main Code]---------------------------------------
Start: high RES
        pulsout clk,TIME
        low RES                    'Reset,==> bit0 on DATA

        For bit_nr = 0 to POS  ' skip unreaded bits
            pulsout clk,time
        next                   ' ===> bit(Pos+1) on DATA

        shiftin DAT,CLK,lsbpre,[mem16\16]
        shiftin DAT,CLK,lsbpre,[mem4\4]

        'gosub shin20'change with both upper shiftin DAT..
'=======================================================
        DEBUG cr,"Serial Number = ", HEX mem4 ,HEX mem16
'=======================================================
end

'-----[Subroutines]-------------------------------------
'
shin20: For bit_nr = POS+1 to POS+16 'read 4 lower digits
            mem16  = mem16 >> 1       'shift right
            mem16.bit15  = in2        'new DATA-bit in
```

```
        pulsout clk,TIME              ' clock the next DATA-bit
   next                               '
For bit_nr = POS+17 to POS+20 'read one upper digit
   mem4 = mem4 >>1
   mem4.bit3 = in2
   pulsout clk,TIME
next
return
```

If you know the position and meaning of each bit, you can read any other chip card formatted by ISO7816. Some bits are also writable. In phone cards, the field of payment is writable—but sorry, only the decrease of the amount of credit, supported by the additional logic of the internal EEPROM!

7.2.16 [BS2]—Home Automation with X-10

The old method of data-networking, through the country-wide power supply with high-voltage lines, is growing to provide a new means of home-automation. It is similar to the modem-transmission of data; a carrier of 120 kHz is overlaid to the line for one or more houses. The ratio of 60 Hz for power to 120 kHz for messages provides an easy selection for transmit and receive. But unlike the use of modems with phone-lines, here special methods are needed to avoid errors with the high-power switching pulses.

So the X-10 code format is now the "De Facto Standard" for Power Line Carrier (P.L.C.) transmissions. Now cheaper devices are available from several firms. Without the need for special information lines, the user can begin home-automation in conventional older buildings with little time and money.

In BASIC Stamp II, the X-10 standard is supported with the statement XOUT. See Chapter 5, the first section, PBASIC Instruction Set, for explanation. Only output signals to a transmission-unit are available.

To demonstrate a simple data-transmission, we use two units for connecting with the power line:

• X-10 Powerhouse™ Power Line Interface PL513 for transmit.
• X-10 Powerhouse™ Lamp Module LM465 for receive.

With the lamp module LM465, you have remote control lamp-switching up to 300 W, ON/OFF, and DIMMING-control. But please note that the lamp module cannot be used with fluorescent lamps, low voltage lamps, or appliances. In Figure 7.80 pictures of both X-10 Powerhouse™ modules are shown.

Figure 7.81 shows the circuit diagram. On the receiving side, all you need to do is plug the unit LM465 into the wall-socket. Sometimes filters on the borders

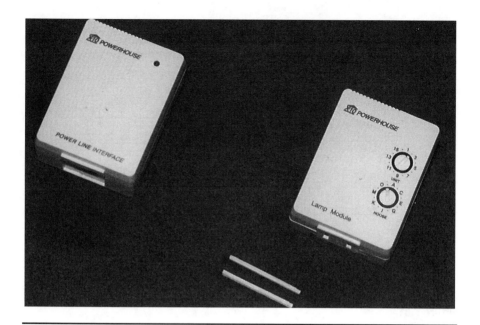

Figure 7.80
Used X-10 Powerhouse™ Modules

of the power-line network are required to limit the damping of transmission level. And filters are needed to avoid influencing ambient parts of the power line.

Now we look at the method for transmitting messages with a home-used power-line, demonstrating with a little program. There is first the "Physical Layer" to transmit a single bit. Next we see how to transmit the full message, with bytes for addressing, to a receiving unit. This is like a "Logical Layer" to take advantage of transmission to the desired unit for processing there the action programmed.

Figure 7.82 shows a 60 Hz period of power-line voltage. In a half-period are embedded three bursts of 120 kHz, each burst with a length of one millisecond. When the zero-crossing point from power-net is achieved (see Fig. 7.81: Line 1 from PL513 to BASIC Stamp II Pin0), the bursts are synchronized to power-line voltage with the timing relationships shown.

A binary 1 is presented by one burst after zero-crossing. To transmit it three times in a half-period is to get coincidence with the zero-crossing points of all three phases in a three-phase power distributing system. If binary 0 is transmitted, there are no bursts.

Note that in a real timing diagram (like oscilloscope-viewing) the bursts are superimposed with power-line voltage, so they sit on the sine wave with a

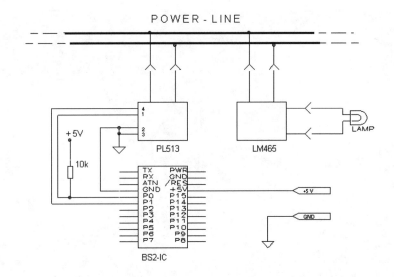

Figure 7.81
Circuit diagram

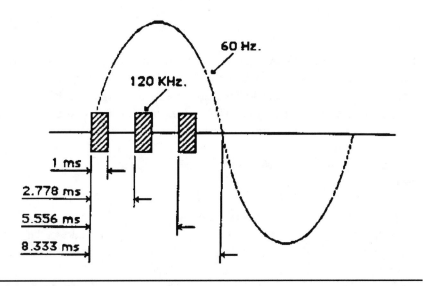

Figure 7.82
Bit transmitting in a half period of 60 Hz

BASIC Stamp

magnitude of 120 V. The magnitude of bursts is less than 1 V (the minimum receiving level is 50 mV$_{ss}$). So Figure 7.82 is only a sketch. We remark that we need a half-period of 8.333 ms to transmit one bit. It's a very slow transmission compared to usual transmission channels. But keep in mind that it is only for home-automation; with no required very quick actions, it is simple to avoid errors with switching pulses.

A complete code transmission contains eleven cycles of power line. The included parts are

- Start-Code, like an "attention" signal—two cycles.
- House-Code to address a unit of some receivers in a group—four cycles. Houses are addressed by one of the characters A, B . . . P.
- Number-Code (Unit Code) to address a special receiver—five cycles or give a Function-Code (for example OFF, ON).

It will be addressed to 16 units or activated by one of 15 function-codes.

The cycles follow each other as shown in Tables 7.30 and 7.31.

To get supplemental security for an error-free transmission, each bit, stored in one half-period of power-line, is followed by the transmission of the inverted value of the same bit in the next half-period of power-line. So a transmission of one bit is doubled within a full period of power-line. Figure 7.83 shows in a symbolic form the transmission of key-number "1" for House-Code "A," for example.

The full code table for houses and units (keys)/functions is shown in Table 7.32. With this information, you can understand the following program example for BASIC Stamp II. For further information on handling X-10 units, see the technical notes, attached to the X-10 Powerhouse™ modules.

Table 7.30
Code transmitted for key-numbers (Unit-Code)

Start-Code	House-Code	Number-Code	Start-Code	House-Code	Number-Code

Table 7.31
Code transmitted for Function-Code

Start-Code	House-Code	Function-Code	Start-Code	House-Code	Function-Code

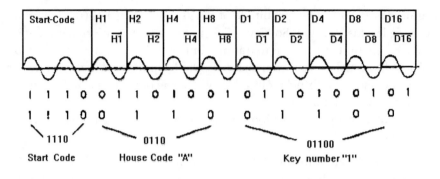

Figure 7.83
Embedded transmission on power-line

Table 7.32
Code table

House-Codes		Key-Numbers		Function-Codes	
A	0110	1	01100	All Units Off	00001
B	1110	2	11100	All Units On	00011
C	0010	3	00100	ON	00101
D	1010	4	10100	OFF	00111
E	0001	5	00010	Status Request	11111
F	1001	6	10010		
G	0101	7	01010		
H	1101	8	11010		
I	0111	9	01110		
J	1111	10	11110		
K	0011	11	00110		
L	1011	12	10110		
M	0000	13	00000		
N	1000	14	10000		
O	0100	15	01000		
P	1100	16	11000		

The usage of the statement XOUT is demonstrated in the following listing, to use with hardware connected as shown in Figure 7.81.

```
' -----[ Title ]-------------------------------------------------
'
' File..        X10out.BS2
' Purpose...    switch ON / OFF a lamp unit
```

```
' Author....      Parallax - Appl.Note 1 /ST. II : X10-control
' Started...
' Updated...


' -----[ Program Description ]-------------------------------------
'  Using  PL513  for transmit, BS2  generates signals for
'  switching  a lamp with receiving module LM465
'
' -----[ Constants ]-----------------------------------------------
          zPin      con       0          'zero crossing detect  from PL513
          mPin      con       1          'modulation control  pin to PL513
          houseA    con       0          ' House code A
          Unit1     con       0          ' Unit Code for unit 1
          uniton    con       %00101 ' Code  lamp ON
          unitoff   con       %00111 ' Code  lamp OFF


' -----[ Variables ]-----------------------------------------------
'       no

' -----[ Initialization ]------------------------------------------
'       no

' -----[ Main Code ]-----------------------------------------------
start:
     pause 1000                           'wait a second
     xout mPin, zPin,[houseA,\Unit1]      'talk to Unit 1
     xout mPin, zPin,[houseA\uniton]      'tell to turn ON
     pause 1000                           'wait a second
     xout mPin, zPin,[houseA\unitoff]     'tell it to turn off
end

' -----[ Subroutines ]---------------------------------------------
' no
```

The main code part of the listing contains three XOUT statements to transmit
X-10 commands by the PL513 unit. First unit "1" in house "A" is addressed. The
unit address is marked for the next two transmissions. So the next two statements include the Function-Codes lamp-ON and lamp-OFF for house "A."
Between the lamp switching, a delay of one second is inserted.

Notice the differences between the codes used in Parallax's application
node and the codes used in the documentation (Rev. 2.4) of the X-10 Power-
house™ modules.

Appendix A
Reserved Words

Reserved Words for BASIC Stamp I

AND	BSAVE	PIN3
B0	BUTTON	PIN4
B1	DEBUG	PIN5
B2	DIR0	PIN6
B3	DIR1	PIN7
B4	DIR2	PINS
B5	DIR3	PORT
B6	DIR4	POT
B7	DIR5	PULSIN
B8	DIR6	PULSOUT
B9	DIR7	PWM
B10	DIRS	RANDOM
B11	EEPROM	READ
B12	END	REM
B13	FOR	RETURN
B14	GOSUB	REVERSE
B15	GOTO	SERIN
BIT0	IF	SEROUT
BIT1	INPUT	SLEEP
BIT2	LET	SOUND
BIT3	LOOKDOWN	STEP
BIT4	LOOKUP	SYMBOL
BIT5	LOW	THEN
BIT6	MAX	TO
BIT7	MIN	TOGGLE
BIT8	NAP	W0
BIT9	NEXT	W1
BIT10	OR	W2
BIT11	OUTPUT	W3
BIT12	PAUSE	W4

BIT13	PIN0	W5
BIT14	PIN1	W6
BIT15	PIN2	WRITE

Reserved Words for BASIC Stamp II

AND	DIR14	IND
BIT0	DIR15	INH
BIT1	DIRA	INL
BIT2	DIRB	INS
BIT3	DIRC	INPUT
BIT4	DIRD	LET
BIT5	DIRH	LOOKDOWN
BIT6	DIRL	LOOKUP
BIT7	DIRS	LOW
BIT8	DTMFOUT	LOWBIT
BIT9	END	LOWBYTE
BIT10	FOR	LOWNIB
BIT11	FREQOUT	MAX
BIT12	GOSUB	MIN
BIT13	GOTO	NAP
BIT14	HIBIT	NIB0
BIT15	HIBYTE	NIB1
BYTE0	HINIB	NIB2
BYTE1	IF	NIB3
BUTTON	IN0	NEXT
CON	IN1	OR
COUNT	IN2	OUT0
DATA	IN3	OUT1
DEBUG	IN4	OUT2
DIR0	IN5	OUT3
DIR1	IN6	OUT4
DIR2	IN7	OUT5
DIR3	IN8	OUT6
DIR4	IN9	OUT7
DIR5	IN10	OUT8
DIR6	IN11	OUT9
DIR7	IN12	OUT10
DIR8	IN13	OUT11
DIR9	IN14	OUT12
DIR10	IN15	OUT13
DIR11	INA	OUT14
DIR12	INB	OUT15
DIR13	INC	OUTA

OUTB	PIN6	SEROUT
OUTC	PIN7	SHIFTIN
OUTD	PINS	SHIFTOUT
OUTH	PORT	SLEEP
OUTL	PULSIN	STEP
OUTS	PULSOUT	STOP
OUTPUT	PWM	THEN
PAUSE	RANDOM	TO
PIN0	RCTIME	TOGGLE
PIN1	READ	VAR
PIN2	REM	WRITE
PIN3	RETURN	XOUT
PIN4	REVERSE	
PIN5	SERIN	

Appendix B
BASIC Stamp—Versions
of Firmware

BASIC Stamp I

V.1.0	January 1993

Version for key costumers—deviated language structure

V. 1.1	January 25, 1993	
PULSIN	-	Time Out Error
SERIN	-	Multiple Qualifier Error
PORT	-	Variable Read Error

V. 1.2	March 30, 1993	
PULSIN	-	Error repaired
PORT	-	Error repaired
SERIN	-	Multiple Qualifier Error

V.1.3	August 18, 1993	
SERIN	-	Error repaired

No further errors known

V.1.4	August 19, 1993	
SEROUT	-	Better Timing

No further errors are known—firmware tested completely

BASIC Stamp II

V.1.0	1995

No errors are known

Appendix C
HAYES Commands—The Most
Important Modem Instructions

+++	Change from transparent mode to command mode
A/	Repeat last command line
ATD	Dial the following dial string
ATDS=n	Dial the stored dial string number # n
ATEn	Command echo on or off
ATH	Hang Up
ATLn	Speaker volume
ATMn	Speaker on or off
ATOn	Change from command mode to transparent mode
ATQn	Feedback messages allow or avoid
ATSn=x	Write value x to control register n
ATSn?	Read value from control register n
ATXn	Modem behavior during connecting
ATZn	Initialize modem
AT&Cn	DCD options
AT&Dn	DTR options
AT&Fn	Initialize modem with factory configuration
AT&H	Help, display of all commands with their definitions
AT&Rn	CTS options
AT&Sn	DSR options
AT&Tn	Test loop
AT&V	Display of actual configuration
AT\En	Data echo in normal operation on or off
AT\Gn	Data flow control on modem side
AT\Jn	Data transfer speed adapt
AT\Xn	Xon/Xoff execute

Allowed Characters in Dial String after ATD

0 . . . 9	Numbers of the phone number to be dialed. Hyphen and space will be ignored.
,	Pause. Length is defined in control register S8 (Standard 2 sec).
P	Pulse dial (set prior phone number—ATDP 123 45 67).
T	Tone dial (set prior phone number—ATDT 123 45 67).

Appendix D
Product Sources

BASIC Stamps

Parallax, Inc.
3805 Atherton Road, Suite 102
Rocklin, CA 95677

(916) 624 8333

(916) 624 8003 [fax]

http://www.paralaxinc.com
PARALLAXINC.COM

Microcontroller PIC16Cxx and Serial EEPROM

Microchip Technology, Inc.
2355 West Chandler Blvd.
Chandler, AZ 85224-6199

(602) 786 7200

(602) 899 7277 [fax]

http://www.microchip.com

ICs for Interfaces and Peripherals

Dallas Semiconductor Corporation
4401 South Beltwood Parkway
Dallas, TX 75244

(214) 450-0400

(214) 450-3715 [fax]

http://www.dalsemi.com/

✉ National Semiconductor Corporation
2900 Semiconductor Drive
P.O. Box 58090
Santa Clara, CA 95052-8090

☎ (910) 721 5000

💻 http://www.nsc.com

✉ Texas Instruments, Inc.
7839 Churchill Way
Dallas, TX 78759

☎ (512) 250 7655

💻 http://www.ti.com

✉ Analog Devices, Inc.
One Technology Way
P.O. Box 9106
Norwood, MA 02062-9106

☎ (617) 329 4700

📠 (617) 326 8703 [fax]

💻 http://www.analog.com

X-10 Home Automation

✉ X-10 (USA), Inc.
91 Ruckman Rd., Box 420
Closter, NJ 07624

☎ (201) 784 9700

📠 (201) 784 9464 [fax]

💻 http://www.hometeam.com/x10/

Distributors

✉ Digi-Key Corporation
701 Brooks Avenue South
Thief River Falls, MN 56701-0677

☎ (218) 681 6674

📠 (218) 681 3380 [fax]

💻 http://www.digikey.com/

Jameco Electronics
1355 Shoreway Rd.
Belmont, CA 94002

 (800) 831 4242 (toll free)
(415) 592 8097

 (800) 237 6948 (toll free) [fax]
(415) 592 2503

http://www.jameco.com

Index